Annals of Mathematics Studies

Number 119

# Calculus on
# Heisenberg Manifolds

by

## Richard Beals and Peter Greiner

PRINCETON UNIVERSITY PRESS

———

PRINCETON, NEW JERSEY
1988

Clothbound editions of Princeton University Press
books are printed on acid-free paper, and binding
materials are chosen for strength and durability.
Paperbacks, while satisfactory for personal collec-
tions, are not usually suitable for library rebinding

Printed in the United States of America
by Princeton University Press, 41 William Street
Princeton, New Jersey

**Library of Congress Cataloging-in-Publication Data**

Beals, Richard, 1938-
    Calculus on Heisenberg manifolds / by Richard Beals and Peter
Greiner.
        p.    cm. – (The Annals of mathematics studies ; 119)
    Bibliography: p.
    Includes index.
    1. Hypoelliptic operators. 2. Calculus. 3. Differentiable
manifolds.    I. Greiner, P. C. (Peter Charles), 1938-    .
II. Title. III. Title: Heisenberg manifolds. IV. Series: Annals of
mathematics studies ; no. 119.
QA329.42.B42 1988                                    88-9939
515.7'242–dc19                                       CIP

ISBN 0-691-08500-5 (alk. paper)
ISBN 0-691-08501-3 (pbk.)

To our parents

Alice Beals

Anthony and Ildiko Greiner

# Contents

# Preface

We study certain hypoelliptic second order partial differential operators. The motivating example, which we treat in detail, is the Kohn operator $\Box_b$ on a $CR$ manifold. A full asymptotic calculus is developed for an algebra of pseudodifferential operators which includes a parametrix for $\Box_b$.

Our intention has been to develop a calculus which is as explicit as possible. We calculate the leading terms of parametrices in the hypoelliptic case and the leading terms of partial parametrices and of Cauchy–Szegö projections in certain nonhypoelliptic cases. Computation of further terms involves computing convolutions of homogeneous functions on certain two-step nilpotent groups; we develop a calculus for such convolutions.

We have attempted to make the exposition concrete, reasonably self-contained, and accessible to readers interested either in several complex variables or in partial differential equations.

It is with pleasure that we take this opportunity to express our appreciation to Annette Yu for the excellent job she did in typing the original manuscript.

New Haven, Conn. U.S.A.
Toronto, Ont. Canada

July 1985

Calculus on Heisenberg Manifolds

# Introduction

The operator $\Box_b$ is a second order partial differential operator associated to the $\bar{\partial}_b$-complex. It is a natural object of intrinsic interest, and also a prototype of a general class of operators and a test case for various theories and constructions. In these three ways (and others) it is similar to the Laplace–Beltrami operator $\Delta$ associated to the De Rham complex on a Riemannian manifold. Our goal has been to develop an operator calculus which bears the same relation to $\Box_b$ as the classical (elliptic) pseudodifferential operator calculus bears to $\Delta$ and which is as explicit as possible both in its form and in the associated methods of computation.

As a framework for discussing the contents of this monograph and the history of $\Box_b$, we first discuss $\Delta$. In local coordinates $\Delta$ has the form

$$(1) \qquad \Delta = \sum a_{jk}(x) D_j D_k + \text{lower order}, \quad D_j = \frac{1}{i}\frac{\partial}{\partial x_j}.$$

In what we call the *kernel method* one seeks to invert $\Delta$ modulo $C^\infty$ functions as an integral operator $Q$ with kernel $k$,

$$(2) \qquad Qu(x) = \int k(x, x - y) u(y) \, dy.$$

Let $\Delta^x$ denote the constant coefficient operator

$$(3) \qquad \Delta^x u(z) = \sum a_{jk}(x) \frac{1}{i}\frac{\partial}{\partial z_j} \frac{1}{i}\frac{\partial}{\partial z_k} u(z).$$

Then $\Delta^x$ has a fundamental solution with convolution kernel $k^x$. Let $Q_0$ be the operator with kernel $k_0$, where $k_0(x, x - y) = k^x(x - y)$. One can show that $\Delta Q_0 = I - R$ where $R$ has a less singular kernel. Setting

$$(4) \qquad Q_N = \sum_{j=0}^{N} Q_0 R^j$$

one has $\Delta Q_N = I - R^{n+1}$, and $R^{N+1}$ has kernel which is $C^m$ in all variables if $N \geq N(m)$. All the local Hölder and $L^p$ regularity results for $\Delta$ can be deduced from this construction. For example, if $\Delta u$ belongs locally to $L^p$, where $1 < p < \infty$, then every second order derivative of $u$ belongs locally to $L^p$.

This kernel method gives the kernel $k$ as an asymptotic sum

(5) $$k \sim k_0 + k_1 + k_2 + \cdots$$

where $k_0$ is known and the $k_j$ are successively less singular. Here we can take $k_j$ to be the kernel of $Q_0 R^j$, but it is difficult to establish directly what form the $k_j$ take.

In the *symbol method* one looks instead for the *symbol*: the function (or distribution $q(x, \cdot)$) such that

(6) $$Qu(x) = \int_{\mathbf{R}^n} e^{ix\cdot\xi} q(x,\xi)\hat{u}(\xi)\frac{d\xi}{(2\pi)^n}.$$

Thus $q(x, \cdot)$ is the partial Fourier transform

(7) $$q(x,\xi) = \int e^{-i\xi\cdot z} k(x,z)dz.$$

Symbols are manipulated much more readily than kernels. The operator $\Delta$ itself has symbol

(8) $$\sigma(\Delta) = \sum a_{jk}(x)\xi_j\xi_k + \text{lower order}.$$

According to the symbol calculus for classical pseudodifferential operators the symbol of the composition $Q'Q''$ of operators having symbols $q'$ and $q''$ is asymptotically

(9) $$q' \circ q'' \sim \sum_{\alpha} \frac{1}{\alpha!} i^{\alpha} (D_{\xi}^{\alpha} q')(D_x^{\alpha} q'').$$

One seeks an expansion of the symbol $q$ of $\Delta^{-1}$:

(10) $$q \sim q_0 + q_1 + q_2 + \cdots$$

where $q_j(x, \cdot)$ is homogeneous of degree $-2 - j$ in the $\xi$ variables. Then necessarily $q_0(x,\xi) = [\sum a_{jk}(x)\xi_j\xi_k]^{-1}$ and the remaining terms $q_j$ are easily computed by recursion.

Pseudodifferential methods involve the Fourier transform, so they are intrinsically $L^2$ theories. Nevertheless in this classical case (10) immediately gives a kernel expansion of the type (5). Here $k_j$ is determined from $q_j$ and it can easily be shown to have the form

(11) $$k_j(x,z) = k_j'(x,z) + p_j(x,z) \log |z|$$

where $k'_j(x, \cdot)$ is homogeneous of degree $2 - n + j$ and $p_j(x, \cdot)$ is a polynomial homogeneous of degree $2 - n + j$. One then recovers the Hölder and $L^p$ regularity.

The symbol method can be adapted to study complex powers $\Delta^s$ or the heat operator $\frac{\partial}{\partial t} + \Delta$, giving enough information on the kernels of these operators to allow one to connect local and global geometric invariants.

## Contents of Chapters 1-4

In Chapter 1 we introduce a class of second order partial differential operators $P$. This class includes both $\square_b$ and the heat operator $\frac{\partial}{\partial t} + \Delta$. Associated to such an operator $P$ on a manifold $M$ is a codimension 1 subbundle $\mathcal{V}$ of the tangent bundle $TM$. In turn we associate to $\mathcal{V}$ and to a point $y \in U$, where $U$ is a coordinate neighborhood identified with a subset of $\mathbf{R}^{d+1}$, a group structure on $\mathbf{R}^{d+1}$ having $y$ as identity element. This group is isomorphic to one of the groups $H_m \times \mathbf{R}^{d-2m}$, where $H_m$ is the $(2m + 1)$-dimensional Heisenberg group; $m = m(y)$ may vary with $y$. There is then a left invariant operator $P^y$ which is a best approximation to $P$ at the point $y$. This "*model operator*" $P^y$ plays the same role for $P$ as the translation invariant operator $\Delta^x$ plays for $\Delta$.

A basic concern is whether $P$ is "*hypoelliptic with loss of one derivative*". This means that if $Pu$ is locally in $L^2$, then each *first* derivative of $u$ is locally in $L^2$, a loss of one derivative relative to the result for $\Delta$. If so, one wants to construct a *parametrix* $Q$ for $P$ (an inverse modulo $C^\infty$). We show in §2 that the model operator $P^y$ satisfies the basic estimate for hypoellipticity with loss of one derivative if and only if it is invertible.

Chapter 2 is devoted to calculating the inverse $Q^y$ of the model operator $P^y$ whenever it exists. Both the symbol and the kernel of $Q^y$ are found explicitly. In appropriate noninvertible cases we calculate the symbol and the kernel of a partial inverse and of the projection onto the kernel of $P^y$.

Chapter 3 begins with a summary of results from the classical theory of pseudodifferential operators. We then introduce the algebra of $\mathcal{V}$-*operators* and develop its calculus in somewhat greater detail than in our paper [1]. As an application we show that an operator $P$ of the type considered in Chapter 1 is hypoelliptic with loss of one derivative if and only if each of the model operators $P^y$ is invertible. If so, the inverses $Q^y$ give the principal term of a full parametrix $Q$ which is a $\mathcal{V}$-operator. This implies in particular that the kernel of $Q$ has an asymptotic expansion in local coordinates

$$(12) \qquad k(x,y) \sim \sum_{j=-d}^{\infty} f_j(x, -\psi_x(y)) + \sum_{j=0}^{\infty} p_j(x, -\psi_x(y)) \log \|\psi_x(y)\|.$$

Here $\psi_x : U \to \mathbf{R}^{d+1}$ is a known coordinate map, $f_j(x, \cdot)$ is a function with parabolic homogeneity of degree $j$, $p_j(x, \cdot)$ is a polynomial with parabolic homogeneity of degree $j$, $\| \quad \|$ denotes a parabolic homogeneous norm, and the principal term $f_{-d}$ is known. We also discuss the global $L^2$ theory of $\mathcal{V}$-operators on a compact manifold.

The motivating example $\square_b$ is treated in detail in Chapter 4. We review the definitions and calculations for the $\overline{\partial}_b$-complex and for $\square_b$ on $(0, q)$-forms. The results of Chapters 1-3 are applied to get a parametrix for $\square_b$ under the hypoellipticity condition $Y(q)$. In certain nonhypoelliptic cases we obtain a partial parametrix and a formal projection to $\ker(\square_b)$. In the $L^2$ theory we show that the exact partial inverse and the Cauchy–Szegö projection are $\mathcal{V}$-operators and agree to infinite order with the operators we obtained. Therefore all these operators have kernels with expansions of the type (12) with leading terms of the appropriate degree.

We work throughout computationally in local coordinates. As in the elliptic case, the principal term of the symbol of a $\mathcal{V}$-operator has an invariant interpretation, but as in the elliptic case the full asymptotic expansion has meaning only with respect to given local coordinates.

## Some History of Symbols, Kernels and $\square_b$

The classical (elliptic) pseudodifferential calculus has its roots in work by Giraud, Mikhlin, and Calderón and Zygmund on singular integral operators; see Seeley's article [3]. A full asymptotic calculus was developed independently by Kohn and Nirenberg [1], Seeley [1], and Unterberger and Bokobza [1]. Applications to global geometric questions via $\Delta^s$ and $\frac{\partial}{\partial t} + \Delta$ were given by Seeley [2], McKean and Singer [1], Greiner [1], Patodi [1], and Gilkey [1].

Hörmander [1] introduced classes of nonelliptic pseudodifferential operators of *type* $(\rho, \delta)$. When $0 \leq \delta < \rho$ there is a full asymptotic calculus; indeed the order of the terms in the asymptotic expansion (9) still approaches $-\infty$ as $|\alpha| \to +\infty$. Beals and Fefferman [1] introduced more general symbol classes and showed that the formal expansion (9) can be useful even for some examples of type $(\frac{1}{2}, \frac{1}{2})$ where the full asymptotic calculus fails;

a key tool is the $L^2$-boundedness result for operators of type $(\frac{1}{2}, \frac{1}{2})$ due to Calderón and Vaillancourt [1].

Kohn and Rossi [1] showed that study of the Dolbeault complex on a complex manifold $X$ with boundary leads naturally to the analogous $\bar{\partial}_b$-complex on the boundary. They used the $\bar{\partial}$-Neumann problem on $X$ to obtain the basic $L^2$ estimates and thus the $L^2$ and $C^\infty$ regularity. See Folland and Kohn [1] for a complete exposition, including the connections with complex geometry and function theory. The survey article by M. Beals, Fefferman, and Grossman [1] contains much information on these and related topics.

Kohn [1] initiated the intrinsic study of the $\bar{\partial}_b$-complex, defining what is now called a $CR$-manifold. He introduced the associated second order operator $\Box_b$ and proved the basic $L^2$ estimates directly.

The operator $\Box_b$ is a naturally occurring operator which is (sometimes) hypoelliptic but is not of "constant strength," so that methods developed for the study of general constant coefficient operators cannot account for its properties. The parametrix turns out to be of type $(\frac{1}{2}, \frac{1}{2})$, so that the pseudodifferential techniques in Hörmander [1] also break down. Thus $\Box_b$, like the closely related unsolvable operator of H. Lewy [1], [2] became a starting point and test case for the study of general operators with variable coefficients.

In discussing the literature we restrict ourselves here to papers which gave new information about $\Box_b$ itself, and we discuss them only in that connection. For related results on hypoelliptic operators with multiple characteristics see Grushin [1], [2], Boutet de Monvel and Treves [1], Menikoff [1], Hörmander [2], Grigis [1], and Boutet de Monvel, Grigis and Helffer [1]. Also we do not consider Gevrey regularity of $\Box_b$, proved by Tartakoff [1], nor analytic hypoellipticity, which was proved by Tartakoff [2], [3] and Treves [1].

Kohn's $L^2$ estimates [1] give the basic $L^2$ and $C^\infty$ regularity for $\Box_b$ on $(0, q)$-forms under condition $Y(q)$. For some values of $q$ this condition is compatible with degeneracy of the Levi form associated to the $CR$ manifold $M$. Of themselves, these estimates do not appear to give any information about the existence and form of a parametrix, nor about regularity in other function spaces (although see the discussion to follow).

The theorem of Egorov [1] allows one "microlocally" to bring $\Box_b$ into very special form by conjugating with Fourier integral operators. Boutet de Monvel [1] obtained this special form and exploited it to construct a parametrix of type $(\frac{1}{2}, \frac{1}{2})$ for $\Box_b$ under condition $Y(q)$, assuming a non-degenerate Levi form. Sjöstrand [1] also used Fourier integral transformations to obtain a parametrix without the nondegeneracy assumption, but did not determine its type. Beals [1] showed that Sjöstrand's parametrix is necessarily also of type $(\frac{1}{2}, \frac{1}{2})$. Unfortunately, the Fourier integral transformations destroy all useful information about the kernel of the parametrix and allow only $L^2$ and $C^\infty$ regularity information.

Folland and Stein [1] introduced the use of the Heisenberg group as a good pointwise model for a $CR$ manifold with nondegenerate Levi form. Extending the construction of Folland [1], they determined kernels for the inverses of model operators on the group, which gave them explicitly the principal term of a parametrix for $\Box_b$ on a manifold with "Levi metric." This knowledge of the kernel allowed them to deduce the first Hölder and $L^p$ regularity results for $\Box_b$.

A rather different kernel approach was developed by Rothschild and Stein [1]. They obtained $L^p$ and Hölder regularity results without restricting to a nondegenerate Levi form and a Levi metric, although they did not determine explicitly the principal term of a parametrix. As in (4) their construction gives an implicit parametrix to any finite degree of smoothing. Rothschild and Tartakoff [1] refined this construction and obtained a full parametrix.

Greiner and Stein [2] showed how to use this information about $\Box_b$ to obtain precise regularity results for the $\bar{\partial}$-Neumann problem on a strictly pseudoconvex domain.

Back on the symbol side, Nagel and Stein [1], [2] first showed how to obtain optimal $L^p$ and Hölder results for nonclassical pseudodifferential operators via information on the symbols. In particular they were able to rederive the results of Rothschild and Stein for $\Box_b$. These ideas were developed in somewhat different form for general pseudodifferential operators by Beals [2], [3]. It is shown in Beals [2] that one can deduce the existence of a parametrix and enough information about its kernel to establish the $L^p$ and Hölder regularity directly from Kohn's $L^2$ estimates for $\Box_b$ and the form of $\Box_b$ itself. The pseudodifferential operators considered by these authors form algebras under composition, but without an asymptotic composition law.

When the Levi form is nondegenerate but condition $Y(q)$ fails, one is interested in the Cauchy–Szegö projection onto the kernel of $\Box_b$. Motivated by results of Fefferman [1] about the Bergman kernel, Boutet de Monvel and Sjöstrand [1] determined an asymptotic expansion for the Cauchy–Szegö projection in the case of functions on a strictly pseudoconvex boundary.

These results were used by Phong and Stein [1] to derive regularity properties of the Cauchy-Szegö projection for functions. Kerzman and Stein [1] showed how the Cauchy–Szegö kernel is related to the Henkin–Ramirez kernel.

What is missing? Some of the problems pointed out by Folland and Stein at the end of [1] remained open. One would like to calculate explicitly the principal term of the parametrix of $\Box_b$ in the general case. One would like a full asymptotic description of the kernel analogous to (5), (11), and a full asymptotic symbol calculus analogous to (9). Such a calculus would be of interest for its own sake and as a model for other calculi going beyond the classical abelian symbol calculus. It might also allow a corresponding development of the geometric theory: the connection beween local and global geometric invariants.

Calculi of this kind have now been developed independently by the authors [1] and by Taylor [2]. Dynin [1], [2] had proposed a calculus of variable-coefficient convolution operators on the Heisenberg group and had noted that this gives a local model for any *contact manifold*. Jerison [1] considered such operators from the kernel side and established a composition theorem modulo a term of lower order. Taylor [2] develops a full asymptotic calculus from the symbol side, using operator-valued symbols and the group Fourier transform. Principal symbols are not computed explicitly. Because of the contact manifold – Heisenberg group setting, applications to $\Box_b$ are restricted to the case of nondegenerate Levi form.

The approach developed by the authors applies to more general manifolds. We have been uncertain of the best terminology and therefore have used the term *"Heisenberg manifold"* in the title. Our approach is modelled more closely on the classical pseudodifferential calculus. As noted above the classical calculus is based on pointwise approximation of an operator $P$ by freezing the coefficients of the principal part. This procedure fails for $\Box_b$. Folland and Stein noted that a possible alternative is to model $P$ at a point by an operator which is left invariant on a certain group. We carry out such a procedure. As we mentioned, the group structure is allowed to vary from point to point. In particular we can accommodate a degenerate Levi form.

In the classical pseudodifferential calculus the terms in the asymptotic expansion (9) are pointwise products. On the kernel side, the pointwise product is enclidean convolution. Our calculus has the same general form, but involves convolution in the group associated to a point.

These methods extend to operators such as the $\Box_b$ heat operator $\frac{\partial}{\partial t} + \Box_b$. Under condition $Y(q)$ the kernel of $\exp(-t\Box_b)$ has an asymptotic expansion as $t \to 0+$ and one can obtain results analogous to the Riemannian case. For these questions, see Beals, Greiner and Stanton [1], [2] and also Stanton and Tartakoff [1], Taylor [2]. Combining these techniques with the methods of Greiner and Stein [2], Beals and Stanton [1] have obtained analogous results for the heat equation associated to the $\overline{\partial}$-Neumann problem. Recently Beals, Greiner, and Stanton [3] used the calculus developed here to obtain regularity results for the Neumann operator and for the Kohn solution to the $\overline{\partial}$-Neumann problem for $(0, q)$-forms under condition $Z(q)$. These results extend those of Greiner and Stein [2] and Chang [1] for $(0, 1)$-forms on strictly pseudoconvex domains.

# CHAPTER 1

# The Model Operators

§1. *Differential Operators and Their Models*

Let $M$ be a smooth manifold of dimension $d + 1$. We consider a second order differential operator $P$ which is assumed (at least locally) to have the form

$$(1.1) \qquad P = -\sum_{j,k=1}^{d} g_{jk} Y_j Y_k - iT + c$$

where the $Y_j$ are real vector fields, $T$ is a complex vector field, the $g_{jk}$ are real functions, and $c$ is a complex function, all varying smoothly, i.e. in a $C^\infty$ fashion on $M$. We assume throughout that

$(1.2)$     *the $Y_j$ are pointwise linearly independent and the matrix $(g_{jk})$ is pointwise positive definite.*

Let $(h_{jk})$ be the unique positive definite square root of $(g_{jk})$ and set

$$(1.3) \qquad X_j = \sum_{k=1}^{d} h_{jk} Y_k.$$

Then (1.1) may be rewritten as

$$(1.4) \qquad P = -\sum_{j=1}^{d} X_j^2 - iT_1 + c_1$$

where $T_1$ is a complex vector field and $c_1$ a complex function. Assumption (1.2) becomes simply

$(1.5)$     *the $X_j$ are pointwise linearly independent.*

Locally we may choose a vector field $X_0$ so that

$(1.6)$     $X_0, X_1, \ldots, X_d$ *are pointwise linearly independent.*

Again (1.4) may be rewritten as

$$(1.7) \qquad P = -\sum_{j=1}^{d} X_j^2 - i\lambda X_0 + \sum_{j=1}^{d} \gamma_j X_j + c_1$$

where $\lambda$ and the $\gamma_j$ are complex-valued functions.

We work now in a coordinate neighborhood $U$, which we identify with an open subset of $\mathbf{R}^{d+1}$, considered as an affine space. This identification having been made, we want to associate to $P$ and to any point $y \in U$ a *model operator* $P^y$. The philosophy is that the model operator should be as simple as possible, to facilitate calculation, but should contain the relevant information. (In the elliptic case this procedure is classical: the model operator is the homogeneous constant coefficient operator which is the principal part of $P$ at $y$, i.e. the procedure is to "freeze the coefficients".) We begin with model vector fields associated to the $X_j$ and the point $y$. There are unique affine coordinates $x_0, \ldots, x_d$ such that

$$(1.8) \qquad \begin{cases} x_j(y) = 0, \quad 0 \le j \le d; \\[2mm] X_j = \frac{\partial}{\partial x_j} + \frac{1}{2}\sum_{k=0}^{d} \beta_{jk}(x)\frac{\partial}{\partial x_k} \quad \text{with} \quad \beta_{jk}(y) = 0, x \in U. \end{cases}$$

(1.9) DEFINITION: *The model vector fields associated to a point $y \in U$ are those which are expressed in the affine coordinates (1.8) as*

$$(1.10) \qquad \begin{aligned} X_0^y &= \frac{\partial}{\partial x_0}, \\[2mm] X_j^y &= \frac{\partial}{\partial x_k} + \frac{1}{2}\sum_{k=1}^{d} b_{jk}x_k\frac{\partial}{\partial x_0}, \quad 1 \le j \le d, \end{aligned}$$

*where the $b_{jk}$ are constants, given by*

$$(1.11) \qquad b_{jk} = \frac{\partial}{\partial x_k}\beta_{j0}(y).$$

(1.12) DEFINITION: *The model operator associated to the operator $P$ given by (1.7) and to the point $y \in U$ is the operator expressed in terms of the model vector fields by*

$$(1.13) \qquad P^y = -\sum_{j=1}^{d}(X_j^y)^2 - i\lambda(y)X_0^y.$$

The model operator and the model vector fields are left invariant with respect to an appropriate group structure in the affine space containing $U$.

(1.14) DEFINITION: *The group structure associated to a point $y \in U$ is given by a translation, which is expressed in terms of the coordinates (1.8) by*

(1.15)
$$\begin{cases} (x \cdot z)_0 = x_0 + z_0 + \frac{1}{2} \sum_{j,k=1}^{d} b_{jk} x_k z_j, \\\\ (x \cdot z)_j = x_j + z_j, \quad 1 \le j \le d, \end{cases}$$

*where the $b_{jk}$ are the constants (1.11).*

It is an easy calculation to show that (1.15) does indeed define a group operation, and that the group is either abelian or a two step nilpotent Lie group.

(1.16) PROPOSITION: *The model vector fields $X_j^y$ and, hence, the model operator $P^y$ are left invariant with respect to the group structure associated to $y$.*

Proof. We note that left-translations commute with right-translations. Hence, it suffices to show that the $X_j^y$ are infinitesimal right-translations. If $e_0, \ldots, e_d$ are the standard basis vectors in the coordinatization (1.8), then clearly $\mathbf{R}e_j$ is a one-parameter subgroup and

(1.17)
$$\frac{d}{dt} f(x \cdot te_j) \mid_{t=0} = X_j f(x).$$

We shall simplify the model operator. The effect of a *quadratic* change of coordinates of the form

(1.18)
$$x_0' = x_0 + \frac{1}{2} \sum_{j,k=1}^{d} q_{jk} x_j x_k,$$

(1.19)
$$x_j' = x_j, \quad 1 \le j \le d,$$

is to convert the matrix $(b_{jk})$ to

(1.20)
$$(b_{jk}') = (b_{jk} + q_{jk} + q_{kj}).$$

In particular, choosing

(1.21)
$$q_{jk} = q_{kj} = -\frac{1}{4}(b_{jk} + b_{kj}),$$

we may arrange for $(b_{jk}')$ to be skew-symmetric, $(c_{jk})$, where

(1.22)
$$c_{jk} = \frac{1}{2}(b_{jk} - b_{kj}).$$

We note that (1.18) and (1.19) is an isomorphism of groups. We also note that (1.8) and (1.11) show that $\frac{1}{2}(b_{jk} - b_{kj})$ is the $\partial/\partial x_0$ component of the commutator $[X_k, X_j]$ at $y$. Thus our skew-symmetric matrix $c_{jk}$ is uniquely determined by

(1.23) $\qquad [X_k, X_j](y) - c_{jk}(y)X_0(y) \in \text{Span}\{X_j(y) : j > 0\}.$

(1.24) DEFINITION: *We say that the model operator*

$$(1.25) \qquad P^y = -\sum_{j=1}^{d}\left(\frac{\partial}{\partial x_j} + \frac{1}{2}\sum_{k=1}^{d} c_{jk}(y)x_k \frac{\partial}{\partial x_0}\right)^2 - i\lambda(y)\frac{\partial}{\partial x_0}$$

*is in skew-symmetric form if $C = (c_{jk})$ is a skew-symmetric matrix.*

Next let $R = (r_{jk})$ denote an orthogonal matrix such that $A = R^t C R$ is in normal form, i.e.

$$
\begin{array}{ccc}
n & n & d-2n
\end{array}
$$

(1.26) $\qquad A = \begin{pmatrix} 0 & -a & 0 \\ a & 0 & 0 \\ 0 & 0 & 0 \end{pmatrix},$

where

(1.27) $\qquad a = \begin{pmatrix} a_1 & & 0 \\ & \ddots & \\ 0 & & a_n \end{pmatrix}$

and the $a_j$'s are positive. Changing coordinates

(1.28) $\qquad\qquad\qquad\qquad z_0 = x_0,$

(1.29) $\qquad\qquad\qquad z_j = \sum_{k=1}^{d} r_{kj} x_k, \quad j = 1, \ldots, d,$

we obtain

$$-\sum_{j=1}^{d}\left(\frac{\partial}{\partial x_j}+\frac{1}{2}\sum_{k=1}^{d}c_{jk}x_k\frac{\partial}{\partial x_0}\right)^2$$

$$=-\sum_{j=1}^{d}\left(\sum_{k=1}^{d}r_{jk}\frac{\partial}{\partial z_k}+\frac{1}{2}\sum_{k=1}^{d}\left[\sum_{m,n=1}^{d}r_{jm}a_{mn}r_{kn}\right]x_k\frac{\partial}{\partial z_0}\right)^2$$

$$=-\sum_{j=1}^{d}\left(\sum_{k=1}^{d}r_{jk}\left[\frac{\partial}{\partial z_k}+\frac{1}{2}\sum_{l=1}^{d}a_{kl}z_l\frac{\partial}{\partial z_0}\right]\right)^2$$

(1.30)
$$=-\sum_{j=1}^{d}\left(\sum_{k=1}^{d}r_{jk}\left[\frac{\partial}{\partial z_k}+\frac{1}{2}\sum_{l=1}^{d}a_{kl}z_l\frac{\partial}{\partial z_0}\right]\right)$$

$$\left(\sum_{m=1}^{d}r_{jm}\left[\frac{\partial}{\partial z_m}+\frac{1}{2}\sum_{n=1}^{d}a_{mn}z_n\frac{\partial}{\partial z_0}\right]\right)$$

$$=-\sum_{k,m=1}^{d}\left(\sum_{j=1}^{d}r_{jk}r_{jm}\right)\left[\frac{\partial}{\partial z_k}+\frac{1}{2}\sum_{l=1}^{d}a_{kl}z_l\frac{\partial}{\partial z_0}\right]$$

$$\left[\frac{\partial}{\partial z_m}+\frac{1}{2}\sum_{n=1}^{d}a_{mn}z_n\frac{\partial}{\partial z_0}\right]$$

$$=-\sum_{k=1}^{d}\left(\frac{\partial}{\partial z_k}+\frac{1}{2}\sum_{l=1}^{d}a_{kl}z_l\frac{\partial}{\partial z_0}\right)^2.$$

Thus an orthogonal change of variables brings the model operator into *normal form*.

(1.31) DEFINITION: *We say that the model operator*

(1.32)
$$P^y=-\sum_{j=1}^{d}\left(\frac{\partial}{\partial x_j}+\frac{1}{2}\sum_{k=1}^{d}a_{jk}(y)x_k\frac{\partial}{\partial x_0}\right)^2-i\lambda(y)\frac{\partial}{\partial x_0}$$

*is in normal form if $A=(a_{jk})$ is given by (1.26) and (1.27).*

§2 *Estimates for the Model Operator*

The second order operator $P$ is said to be *hypoelliptic with a loss of one derivative* in $U$ if

$$(2.1) \qquad u \in \mathcal{D}'(U), \quad Pu \in H_{\mathrm{loc}}^s(U) \to u \in H_{\mathrm{loc}}^{s+1}(U), s \in \mathbf{R}.$$

Here $H_{\mathrm{loc}}^s(U)$ denotes the localization of the standard Sobolev space $H^s$. Let $\|u\|_{\mathcal{H}}^2$ denote the norm

$$(2.2) \qquad \|u\|^2 + \sum_{j=0}^d \|X_j^y u\|^2 + \sum_{j,k=1}^d \|X_j^y X_k^y u\|^2, \quad u \in C_c^\infty,$$

where $\|\dots\|$ denotes the $L^2$-norm. Let $\mathcal{H}$ denotethe closure of $C_c^\infty$ in the $\|\dots\|_{\mathcal{H}}$ norm.

(2.3) THEOREM: *The operator $P$ is hypoelliptic with loss of one derivative in $U$ if and only if for each $y \in U$ the model operator $P^y$ is bijective from $\mathcal{H}$ to $L^2$.*

In various guises and in various degrees of generality, this theorem can already be found in the literature, e.g. Hörmander [1]. We shall prove it later, and also show that computing the inverse of the model operator allows us to compute the principal term of a complete asymptotic expansion for the parametrix of $P$. In this section we derive the necessary and sufficient condition for the desired invertibility of $P^y$. We set $Z_+ = \{0, 1, 2, \dots\}$.

(2.4) DEFINITION: *Given $y \in U$, the singular set $\Lambda_y$ is defined as follows. Let $(b_{jk})$ be the matrix with elements (1.11) and let*

$$(2.5) \qquad \pm ia_1, \dots, \pm ia_n, a_j > 0,$$

*be the nonzero eigenvalues of the skew symmetric matrix with elements $(b_{jk} - b_{kj})$, repeated according to multiplicity. Then*

$$(2.6) \qquad \Lambda_y = \mathbf{R} \quad if \quad n = 0;$$

$$(2.7) \qquad \Lambda_y = \left\{ \pm \sum_{j=1}^n (2\alpha_j + 1)a_j : \alpha \in Z_+^n \right\} \quad if \quad 2n = d;$$

$$(2.8) \qquad \Lambda_y = \left\{ \Lambda \in \mathbf{R}, |\lambda| \geq \sum_{j=1}^n a_j \right\} \quad if \quad 2n < d.$$

(2.9) THEOREM: *The model operator $P^y$, given by (1.13) satisfies the a-priori estimate*

(2.10)                     $$\|u\|_{\mathcal{H}} \leq C\|P^y u\|, \quad u \in C_c^\infty,$$

*if and only if $\lambda(y)$ does not belong to the singular set $\Lambda_y$.*

From now on we leave off the superscript $y$ from $X_j^y$, $j = 0, 1, \ldots, d$ and from $P^y$ – i.e. $X_0, X_1, \ldots, X_d$ are given by (1.10) and $P$ by (1.13). First we note that the norm (2.2) is essentially independent of the choice of $X_1, \ldots, X_d$. In fact let $e = (e_{jk})$ denote a nonsingular linear transformation of $\mathbf{R}^d$ and set

(2.11)                            $$Y_0 = X_0,$$

(2.12)                    $$Y_j = \sum_{k=1}^{d} e_{jk} X_k, j = 1, \ldots, d.$$

Then

(2.13)
$$\|u\|_{\mathcal{H},Y}^2 = \|u\|^2 + \sum_{j=0}^{d} \|Y_j u\|^2 + \sum_{j,k=1}^{d} \|Y_j Y_k u\|^2$$

$$\leq C\left( \|u\|^2 + \sum_{j=0}^{d} \|X_j u\|^2 + \sum_{j,k=1}^{d} \|X_j X_k u\|^2 \right).$$

Since $e$ is nonsingular, this shows that

(2.14)                    $$\|u\|_{\mathcal{H}} = \|u\|_{\mathcal{H},X} \approx \|u\|_{\mathcal{H},Y}.$$

An integration by parts yields $\|X_j u\|^2 \leq \|u\| \, \|X_j^2 u\|$, $j = 1, \ldots, d$, so

(2.15)              $$\|u\|_{\mathcal{H}} \approx \|u\| + \|X_0 u\| + \sum_{j,k=1}^{d} \|X_j X_k u\|.$$

Next we show that

(2.16)         $$\|X_j X_k u\|^2 \leq C(\|X_j^2 u\|^2 + \|X_k^2 u\|^2 + \|X_0 u\|^2),$$

$u \in C_c^\infty$. We start with

$$X_j X_k u = X_k X_j u + c X_0 u,$$
$$X_0 X_j u = X_j X_0 u,$$

where $c$ denotes a generic constant. Then integrating by parts we have

$$
\begin{aligned}
(X_jX_ku, X_jX_ku) &\leqq |(X_jX_ku, X_kX_ju)| + |(X_jX_ku, cX_0u)| \\
&\leqq |(X_kX_jX_ku, X_ju)| + |(X_jX_ku, cX_0u)| \\
&\leqq |(X_jX_k^2u, X_ju)| + |(cX_0X_ku, X_ju)| \\
&\quad + |(X_jX_ku, cX_0u)| \\
&\leqq \|X_j^2u\|^2 + \|X_k^2u\|^2 + 2|(cX_0u, X_kX_ju)| \\
&\leqq |X_j^2u\|^2 + \|X_k^2u\|^2 \\
&\quad + \frac{1}{2}\|X_jX_ku\|^2 + 2c\|X_0u\|^2,
\end{aligned}
$$

(2.17)

which proves (2.16). Now (2.15) and (2.16) show that

$$
\text{(2.18)} \qquad \|u\|_{\mathcal{H}} \approx \|u\| + \|X_0u\| + \sum_{j=1}^{d}\|X_j^2u\|, \quad u \in C_c^\infty.
$$

Next we consider the behavior of $X_j, j = 1,\ldots,d$ under an orthogonal change of the variables $(x_1,\ldots,x_d) = x'$. Thus, let $R = (r_{ij}) \in O(d)$ and set

$$
\text{(2.19)} \qquad y_0 = x_0, \quad y' = Rx'.
$$

$$
\text{(2.20)} \qquad B = R^t B' R, \quad \text{where} \quad B = (b_{jk}).
$$

Then $X_0 = Y_0$ and

$$
\begin{aligned}
X_j &= \frac{\partial}{\partial x_j} + \frac{1}{2}\sum_{k=1}^{d} b_{jk}x_k\frac{\partial}{\partial x_0} \\[2mm]
&= \frac{\partial}{\partial x_j} + \frac{1}{2}\sum_{k=1}^{d}\left(\sum_{m,n=1}^{d} r_{mj}b'_{mn}r_{nk}\right)x_k\frac{\partial}{\partial x_0} \\[2mm]
&= \sum_{l=1}^{d} r_{lj}\left(\frac{\partial}{\partial y_l} + \frac{1}{2}\sum_{m=1}^{d} b'_{lm}y_m\frac{\partial}{\partial y_0}\right) \\[2mm]
&= \sum_{l=1}^{d} r_{lj}y_l.
\end{aligned}
$$

(2.21)

Consequently the orthogonal change (2.19) of the variables $x_1, \ldots, x_d$ induces a new set of vector fields $y_1, \ldots, y_d$ where $B$ is replaced by $B'$. (2.14) applies and

$$(2.22) \qquad \|u\|_{\mathcal{H},X} \approx \|u\|_{\mathcal{H},Y}.$$

Proof of Theorem 2.9. As noted in (1.22), by making a quadratic coordinate change we may assume that $B = (b_{jk})$ is skew-symmetric. After an orthogonal change we may assume the model vector fields are in normal form, i.e.

$$(2.23) \qquad \begin{cases} X_0 = \frac{\partial}{\partial x_0}, \\ X_j = \frac{\partial}{\partial x_j} - \frac{1}{2} x_{n+j} a_j \frac{\partial}{\partial x_0}, & 1 \le j \le n, \\ X_{n+j} = \frac{\partial}{\partial x_{n+j}} + \frac{1}{2} x_j a_j \frac{\partial}{\partial x_0}, & 1 \le j \le n, \\ X_k = \frac{\partial}{\partial x_k}, & 2n < k \le d. \end{cases}$$

This is justified by (1.30) and (2.22). It is convenient here to make one more quadratic change of the variables of the kind (1.18) and (1.19) with

$$q_{j,n+j} = q_{n+j,j} = \frac{1}{4} a_j, \quad 1 \le j \le n,$$

and other coefficients zero. After this change we have

$$X_j = \frac{\partial}{\partial x_j}, \quad 0 \le j \le n \quad \text{and} \quad 2n < j \le d,$$

$$X_{n+j} = \frac{\partial}{\partial x_{n+j}} + x_j a_j \frac{\partial}{\partial x_0}, \quad 1 \le j \le n.$$

Now we are ready to derive the estimate (2.10). By (2.18) it suffices to show that

$$(2.24) \qquad \|u\|^2 + \|X_0 u\|^2 + \sum_{j=1}^{d} \|X_j^2 u\|^2 \le C \|Pu\|^2,$$

$u \in C_c^\infty$. First we assume that not all $a_j$'s vanish in (2.23). We take the Fourier transform in the variables $(x_0; x_{n+1}, \ldots, x_{2n}; x_{2n+1}, \ldots, x_d)$ and denote the dual variables by $(\tau; \xi_1, \ldots, \xi_n; \eta_1, \ldots, \eta_{d-2n}) = (\tau, \xi, \eta)$. By Plancherel's theorem (2.24) is equivalent to the following family of inequalities for $(\tau, \xi, \eta) \in \mathbf{R} \times \mathbf{R}^n \times \mathbf{R}^{d-2n}$:

$$\|v\|^2 + \sum_{j=1}^{n} (\|D_j^2 v\|^2 + \|(\xi_j + \tau M_j)^2 v\|^2) + \left( \tau^2 + \sum_{j=1}^{d-2n} \eta_j^4 \right) \|v\|^2$$

$$(2.25)$$

$$\le C \|P_{\tau,\xi,\eta} v\|^2, \quad v \in C_c(\mathbf{R}^n),$$

(c) Finally, if $a_j = 0, j = 1, \ldots, n$, then (2.24) implies the following inequality

$$\tau^2 + \sum_{j=1}^{d} |\eta_j|^4 \leq C(\lambda\tau + |\eta|^2)^2,$$

$\eta \in \mathbf{R}^d$ and $\tau \in \mathbf{R}$. Such a constant $C$, independent of $\eta$ and $\tau$, can be found if and only if $\lambda \notin \mathbf{R}$. Thus we have proved Theorem (2.9).

We note that $P^y$ can be extended continuously to $\mathcal{H}$. Then (2.10) is valid on all of $\mathcal{H}$. Now (2.10) implies that $P^y$ is one-to-one on $\mathcal{H}$ with closed range in $L^2$. Hence we have

(2.39) COROLLARY: $P^y$ *has a bounded left inverse on* $L^2$ *if and only if* $\lambda(y) \notin \Lambda_y$.

# CHAPTER 2

# Inverting the Model Operator

## §3 *Operators, Symbols, Composition and Invariance*

To compute the inverse of a model operator, we shall first compute the *symbol* of the inverse, in the sense of pseudodifferential operators. We begin with a brief introduction to the subject.

Let $V = \{x\}$ be a finite-dimensional real vector space with dual space $V' = \{\xi\}$ and pairing $\langle x, \xi \rangle$ on $V \times V'$. If

$$Q : C^\infty(V) \to C^\infty(V)$$

is a linear differential operator with smooth coefficients, its *symbol* is the function

$$q : V \times V' \to \mathbf{C},$$
$$q(x, \xi) = e_{-\xi}(x)[Qe_\xi](x), \quad e_\xi(x) = e^{i\langle x, \xi \rangle}.$$

Thus

$$(3.1) \qquad\qquad [Qe_\xi](x) = q(x, \xi)e_\xi(x).$$

In coordinates $x_1, \ldots, x_n$ with dual coordinates $\xi_1, \ldots, \xi_n$ such an operator $Q$ has the form

$$Q = \sum_\alpha q_\alpha(x)D^\alpha,$$

$$D^\alpha = D_x^\alpha = \left(\frac{1}{i}\frac{\partial}{\partial x_1}\right)^{\alpha_1} \cdots \left(\frac{1}{i}\frac{\partial}{\partial x_n}\right)^{\alpha_n} = i^{-|\alpha|}\partial_x^\alpha.$$

Then

$$(3.2) \qquad\qquad q(x, \xi) = \sum_\alpha q_\alpha(x)\xi^\alpha.$$

Choose translation invariant measures $dx$ on $V$ and $d\xi$ on $V'$. The Fourier transform and Fourier inversion formulas are

$$(3.3) \qquad\qquad \hat{u}(\xi) = \int_V e_{-\xi}(x)u(x)\,dx,$$

(3.4)                        $$u(x) = c \int_{V'} e_\xi(x)\hat{u}(\xi)\,d\xi,$$

where $u$ belongs to the Schwartz class $\mathcal{S} = \mathcal{S}(V)$. We shall assume the measures to be normalized so that

(3.5)                              $$c = (2\pi)^{-m}$$

when $V$ has dimension $m$. In particular on $V = \mathbf{R}^m$ we take $dx$ to be the standard Lebesque measure.

In view of (3.1) and (3.3), we have for a differential operator $Q$ that

(3.6)        $$Qu(x) = (2\pi)^{-m} \int_{V'} e_\xi(x)q(x,\xi)\hat{u}(\xi)\,d\xi, \quad u \in \mathcal{S}.$$

The identities (3.1), (3.6) establish a bijective correspondence between differential operators $Q$ and functions $q$ which are polynomials in $\xi$. For a more general function $q$ (3.6) may be used to *define* a corresponding operator $Q = q(x, D)$. Once again, $q$ will be called the symbol of $Q$.

(3.7) DEFINITION: $S^\infty$ *is the space of $C^\infty$ functions on $V \times V'$ which satisfy the following estimates in coordinates $x_1, \ldots, x_n$ and dual coordinates $\xi_1, \ldots, \xi_n$: for each compact $K \subset V$ and each pair of multi-indices $\alpha$ and $\beta$ there are constants $C, N$ such that*

(3.8)                    $$\sup_{x \in K} |D_x^\alpha D_\xi^\beta q(x,\xi)| \le C(1 + |\xi|)^N.$$

(3.9) PROPOSITION: *If $q$ is in $S^\infty$, then $Q$ defined by (3.6) maps $\mathcal{S}(V)$ into $C^\infty(V)$.*

Proof. If $u$ is in $\mathcal{S}(V)$, then $\hat{u}$ is of rapid decrease in $\xi$ and so the integral in (3.6) is absolutely convergent and can be differentiated with respect to $x$.

(3.10) PROPOSITION: *Suppose $q_1$ and $q_2$ are in $S^\infty$ and let $Q_1, Q_2$ be the corresponding operators. If $q_1$ is a polynomial in $\xi$ or $q_2$ is a polynomial in $x$, then the composition $Q_1 Q_2$ maps $\mathcal{S}(V)$ to $C^\infty(V)$ and has symbol*

(3.11)                    $$q_1 \circ q_2 = \sum_\alpha \frac{1}{\alpha!} \partial_\xi^\alpha q_1 D_x^\alpha q_2.$$

Proof. Suppose first that $q_1$ is a polynomial in $\xi$. Then $Q_1$ is differential and defined on $C^\infty$, so $Q_1 Q_2 : \mathcal{S} \to C^\infty$. To compute the symbol we use the identity

(3.12)          $$D_x^\alpha(e_\xi(x)q(x,\xi)) = e_\xi(x)(D_x + \xi)^\alpha q(x,\xi).$$

This gives

$$(3.13) \quad Q_1 Q_2 u(x) = (2\pi)^{-m} \int_{V'} e_\xi(x) \left[ \sum_\alpha q_{1\alpha}(x)(D_x + \xi)^\alpha q_2(x, \xi) \right] \hat{u}(\xi) d\xi,$$

where

$$Q_1 = \sum_\alpha q_{1\alpha}(x) D_x^\alpha.$$

The expression in brackets in (3.13) is just (3.11).

Suppose now that $q_2$ is a polynomial in $x$,

$$Q_2(x, \xi) = \sum_\alpha x^\alpha q_{2\alpha}(\xi).$$

Then

$$Q_2(u(x)) = \sum_\alpha x^\alpha Q_{2\alpha} u(x),$$

where $Q_{2\alpha}$ has symbol $q_{2\alpha}$. Under our hypothesis, multiplication by $q_{2\alpha}$ maps $\mathcal{S}(V')$ to itself, and it follows that $Q_2$ maps $\mathcal{S}(V)$ to itself and so $Q_1 Q_2 : \mathcal{S} \to C^\infty$. For the Fourier transform,

$$(Q_2 u)\widehat{\ }(\xi) = \sum_\alpha (-D_\xi)^\alpha [q_{2\alpha}(\xi) \hat{u}(\xi)].$$

Applying $Q_1$ and integrating by parts, we get
(3.14)

$$Q_1 Q_2 u(x) = (2\pi)^{-m} \int_{V'} \sum_\alpha q_{2\alpha}(\xi) D_\xi^\alpha [e_\xi(x) q_1(x, \xi)] \hat{u}(\xi) \, d\xi$$

$$= (2\pi)^{-m} \int_{V'} e_\xi(x) \left[ \sum_\alpha q_{2\alpha}(\xi)(D_\xi + x)^\alpha q_1(x, \xi) \right] \hat{u}(\xi) \, d\xi.$$

Again, the expression in brackets in (3.14) is (3.11).

We consider now the effect of an affine coordinate change.

(3.15) PROPOSITION: *Suppose $Q$ has symbol $q \in S^\infty$ and suppose $\psi : V \to V$ is an invertible affine map. Then the operator*

$$(3.16) \qquad\qquad Q_\psi u(x) = Q(u \circ \psi)(\psi^{-1}(x))$$

*has symbol*

$$(3.17) \qquad\qquad q_\psi(x, \xi) = q(\psi^{-1}(x), \, (d\psi)^t(\xi)).$$

Proof. The map $\psi$ has the form $\psi(x) = x^0 + Ax$, where $x_0 \in V$ is fixed and $A : V \to V$ is linear. In calculating $Q_\psi$ we write

$$x = \psi(x'), \quad \xi = (A^{-1})^t \xi', \quad y = \psi(y').$$

Note that with these notations,

$$\langle x' - y', \xi' \rangle = \langle x - y, \xi \rangle, \quad dy' \, d\xi' = dy \, d\xi.$$

Therefore

$$Q_\psi u(x) = (2\pi)^{-m} \int_{V'} e^{i\langle x', \xi' \rangle} q(x', \xi')(u \circ \psi)\widehat{\phantom{.}}(\xi') d\xi'$$

$$= (2\pi)^{-m} \iint_{V \times V'} e^{i\langle x' - y', \xi' \rangle} q(x', \xi') u(y) dy' \, d\xi'$$

$$= (2\pi)^{-m} \iint_{\psi(V) \times (A^{-1})^t(V')} e^{i\langle x - y, \xi \rangle} q(x', \xi') u(y) dy \, d\xi$$

$$= (2\pi)^{-m} \int_{(A^{-1})^t(V')} e_\xi(x) q(x', \xi') \hat{u}(\xi) d\xi,$$

which gives (3.17).

Suppose now that $V = \mathbf{R}^{d+1}$ with coordinates $x_0, \ldots, x_d$ and with the group structure of §1:

$$(3.18) \qquad \begin{cases} (x \cdot y)_0 = x_0 + y_0 + \frac{1}{2} \sum\limits_{j,k=1}^{d} b_{jk} x_k y_j, \\[2mm] (x \cdot y)_j = x_j + y_j, \quad j > 0. \end{cases}$$

Dropping the superscripts $y$, the corresponding vector fields of (1.10) are

$$(3.19) \qquad \begin{cases} X_0 = \frac{\partial}{\partial x_0}, \\[2mm] X_j = \frac{\partial}{\partial x_j} + \frac{1}{2} \sum\limits_{k=1}^{d} b_{jk} x_k \frac{\partial}{\partial x_0}, \quad j > 0. \end{cases}$$

Clearly, $X_j$ has symbol $i\sigma_j$, where

$$(3.20) \qquad \begin{cases} \sigma_0 = \xi_0, \\[2mm] \sigma_j = \xi_j + \frac{1}{2} \sum\limits_{k=1}^{d} b_{jk} x_k \xi_0, \quad j > 0. \end{cases}$$

As usual, if $Q$ maps functions on $\mathbf{R}^{d+1}$ to functions on $\mathbf{R}^{d+1}$, we say that $Q$ is *left invariant* for the group structure if $Q_\psi = Q$ for every left translation $\psi$, where $Q_\psi$ is defined by (3.16).

(3.21) PROPOSITION: *Suppose $Q$ has symbol $q \in S^\infty$. Then $Q$ is left invariant if and only if there is a function $f \in C^\infty(\mathbf{R}^{d+1})$ such that*

$$(3.22) \qquad q(x, \xi) = f(\sigma_0(x, \xi), \ldots, \sigma_d(x, \xi)), \quad all \quad (x, \xi) \in V \times V'.$$

Proof. Each left translation is an affine map, so Proposition 3.15 applies: left invariance means

$$q(x, \xi) = q(\psi^{-1}(x), (d\psi)^t \xi)$$

for every left translation $\psi$, and conversely. In particular this is true for $\psi_x$, left translation by $x$, and gives

$$(3.23) \qquad\qquad q(x, \xi) = f((d\psi_x)^t \xi)$$

where $f(\eta) = q(0, \eta)$. In particular,

$$(3.24) \qquad\qquad \sigma_j(x, \xi) = f_j((d\psi_x)^t \xi).$$

Taking $x = 0$, we see that $f_j(\eta) = \eta_j$, so (3.24) is

$$(3.25) \qquad\qquad (d\psi_x)^t \xi = (\sigma_0(x, \xi), \ldots, \sigma_d(x, \xi)).$$

Then (3.23) and (3.25) give (3.22). This proves Proposition 3.21.

We turn now to the model operator (1.13). Dropping the $y$-dependence once again,

$$(3.26) \qquad\qquad P_\lambda = -\sum_{j=1}^{d} X_j^2 - i\lambda X_0.$$

Thus $P_\lambda$ has symbol

$$(3.27) \qquad\qquad P_\lambda(x, \xi) = \sum_{j=1}^{d} \sigma_j(x, \xi)^2 + \lambda \sigma_0(x, \xi).$$

Let us assume now that normal coordinates have been chosen so that the model vector fields have the form (2.23), which we repeat here for convenience:

$$(3.28) \qquad \begin{cases} X_0 = \frac{\partial}{\partial x_0}, \\[2mm] X_j = \frac{\partial}{\partial x_j} - \frac{1}{2} a_j x_{n+j} \frac{\partial}{\partial x_0}, \quad 1 \le j \le n, \\[2mm] X_{n+j} = \frac{\partial}{\partial x_{n+j}} + \frac{1}{2} a_j x_j \frac{\partial}{\partial x_0}, \quad 1 \le j \le n, \\[2mm] X_k = \frac{\partial}{\partial x_k}, \quad 2n < k \le d, \end{cases}$$

where the $a_j$'s are positive. It is easy to check that

$$X_j^2 + X_{n+j}^2$$

is invariant under rotation in the $(x_j, x_{n+j})$ plane, so $P_\lambda$ is invariant under each such rotation.

If $P_\lambda$ is invertible, its inverse $Q_\lambda$ is left invariant. Therefore if $Q_\lambda$ has a symbol $q_\lambda$, the symbol is a function of the $\sigma_j$'s. Moreover, $Q_\lambda$ must be invariant under rotations in each $(x_j, x_{n+j}$ plane, $1 \le j \le n$. This means $q_\lambda$ has the form (3.22) and rotation invariance yields

$$(3.29) \qquad \xi_{n+j}\frac{\partial f}{\partial \xi_j} - \xi_j \frac{\partial f}{\partial \xi_{n+j}} = 0, \quad j = 1, \dots, n.$$

With more convenient notation we have

(3.30) PROPOSITION: *If the operator $P_\lambda$ given by (3.26) has a left inverse $Q_\lambda$ with symbol $\sigma(Q_\lambda) \in S^\infty$, then*

$$(3.31) \qquad \sigma(Q_\lambda)(x,\xi) = q_\lambda(\sigma_0(x,\xi), \dots, \sigma_d(x,\xi))$$

*where $q_\lambda$ satisfies the differential equation*

$$(3.32) \qquad \sum_{j=1}^{2n}\left[\xi_j^2 - \frac{1}{4}a_j^2\xi_0^2\left(\frac{\partial}{\partial \xi_j}\right)^2\right]q_\lambda + \sum_{j=2n+1}^{d}\xi_j^2 q_\lambda + \lambda\xi_0 q_\lambda = 1,$$

*where $a_{n+j} = a_j$.*

Proof. If $Q_\lambda$ is a left inverse we use Proposition 3.10 with $Q_1 = Q_\lambda$ and $Q_2 = P_\lambda$ to get

$$(3.33) \qquad 1 = \sum_\alpha \frac{1}{\alpha!}\partial_\xi^\alpha \sigma(Q_\lambda)D_x^\alpha \sigma(P_\lambda).$$

We use (3.22) and evaluate at $x = 0$. Because of (3.29) the terms with $|\alpha| = 1$ vanish and we obtain (3.32).

(3.34) REMARK: We shall show in §4 that $P_\lambda$ has a two-sided inverse $Q_\lambda'$ on test functions, i.e.

$$(3.35) \qquad P_\lambda Q_\lambda' u = Q_\lambda' P_\lambda u = u, \quad u \in \mathcal{S}$$

when $\lambda \notin \Lambda$. It follows that the range of $P_\lambda$, as a mapping from $\mathcal{H}$ to $L^2$, is dense in $L^2$. Consequently the left inverse $Q_\lambda$ is a two-sided inverse and $Q_\lambda$ is the closure of $Q_\lambda'$.

(3.36) REMARK: For simplicity we have assumed our symbols are smooth at $\xi = 0$. This will not actually be the case for the symbol of the inverse of the model operator, although the solution of (3.32) does yield the correct symbol of the inverse. This will be discussed more fully below.

(3.37) REMARK: From now on we shall denote all the symbols of $p_\lambda^{-1}$ by $q_\lambda$, i.e.

$$(3.38) \qquad \sigma(P_\lambda^{-1})(x,\xi) = q_\lambda(\sigma(x,\xi)).$$

Thus the $q_\lambda$ of §4, giving the inverse symbol for $P_\lambda$ when $P_\lambda$ is in normal form is a special case of the $q_\lambda$ of §6, where we find the inverse symbol of $P_\lambda$ when $P_\lambda$ is in general skew-symmetric form. This, in turn is a special case of the $q_\lambda$ of §8 which gives $\sigma(P_\lambda^{-1}(x,\xi))$ for the general model operator $P_\lambda$.

§4 *Inverting $P_\lambda$ in Normal Form: Symbols*

We proceed now to calculate the symbol of the inverse of the model operator

$$(4.1) \qquad P_\lambda = -\sum_{j=1}^{d} X_j^2 - i\lambda X_0$$

where the $X_j$'s are given by (3.28) and we assume $n > 0$. Thus we want to solve the differential equation (3.32), which we write here as

$$(4.2) \qquad (L+\mu)q_\lambda = 1, \quad \mu = \lambda\xi_0 + \sum_{j=2n+1}^{d} \xi_j^2$$

$$(4.3) \qquad Lq_\lambda(\xi) = \sum_{j=1}^{2n}\left[\xi_j^2 - \frac{1}{4}a_j^2\xi_0^2\left(\frac{\partial}{\partial\xi_j}\right)^2\right]q_\lambda.$$

Our derivation of a formula for the solution of (4.2) will be heuristic and formal, since we may check directly at the end that what we have obtained will in fact satisfy (4.2).

The operator $L$ is a sum of one-dimensional Hermite operators, so it is natural to expand in products of the corresponding eigenfunctions. Recall that these eigenfunctions are defined by the identity

$$(4.4) \qquad \sum_{k=0}^{\infty} \frac{t^k}{k!}\psi_k(x) = \exp(2tx - t^2 - \frac{1}{2}x^2), x, t \in \mathbf{R}.$$

As a consequence of (4.4),

(4.5) $$\left[x^2 - \left(\frac{d}{dx}\right)^2\right]\psi_k(x) = (2k+1)\psi_k(x).$$

We assume $\xi_0 \neq 0$ and set

(4.6) $$\psi_\alpha(\xi_1, \ldots, \xi_{2n}) = \prod_{j=1}^{2n} \psi_{\alpha_j}(\beta_j \xi_j), \alpha \in Z_+^{2n},$$

where

(4.7) $$\beta_j = \left(\frac{1}{2}|\xi_0|a_j\right)^{-\frac{1}{2}}.$$

Then (4.5) - (4.7) give

(4.8)) $$L\psi_\alpha = \mu_\alpha \psi_\alpha,$$

where

(4.9) $$\mu_\alpha = \mu_\alpha(\xi_0) = \sum_{j=1}^{2n} \frac{1}{2}a_j|\xi_0|(2\alpha_j + 1).$$

The $\psi_k$'s are an orthogonal basis for $L^2(\mathbf{R})$ (see below), so the $\psi_\alpha$'s are an orthogonal basis for $L^2(\mathbf{R}^{2n})$. We look for $q_\lambda(\xi_1, \ldots, \xi_{2n})$ having the form

(4.10) $$q_\lambda = \sum_\alpha c_\alpha \psi_\alpha, \quad c_\alpha = c_\alpha(\xi_0, \xi_{2n+1}, \ldots, \xi_d).$$

Because of (4.8), equation (4.2) becomes

(4.11) $$\sum_\alpha c_\alpha[\mu_\alpha + \mu]\psi_\alpha \equiv 1.$$

Now fixing $t$ in (4.4) and taking inner products in $L^2(\mathbf{R})$, we have

(4.12) $$(\psi_j, \psi_k) = 2^k k! \pi^{\frac{1}{2}} \delta_{jk},$$
$$(\psi_{2k+1}, 1) = 0, \quad (\psi_{2k}, 1) = (2k)!(2\pi)^{\frac{1}{2}}(k!)^{-1}.$$

Therefore in $L^2(\mathbf{R}^{2n})$,

$$(\psi_\alpha, \psi_\beta) = 2^{|\alpha|}\alpha!\pi^n \delta_{\alpha\beta} \prod_{j=1}^{2n} \beta_j^{-1},$$

(4.13) $$(\psi_\alpha, 1) = 0, \quad \alpha \notin (2Z_+)^{2n},$$

$$(\psi_{2\alpha}, 1) = (2\alpha)!(2\pi)^n(\alpha!)^{-1} \prod_{j=1}^{2n} \beta_j^{-1}, \quad \alpha \in Z_+^{2n}.$$

Taking inner products in (4.11) we obtain

(4.14) $$c_\alpha[\mu_\alpha + \mu](\psi_\alpha, \psi_\alpha) = (1, \psi_\alpha)$$

or

$$c_\alpha = 0, \quad \alpha \notin (2Z_+)^{2n}, \quad c_{2\alpha} = 2^{n-2|\alpha|}[\mu_{2\alpha} + \mu]^{-1}(\alpha!)^{-1}.$$

Thus

(4.15) $$q_\lambda = \sum_\alpha \frac{2^{n-2|\alpha|}}{\alpha!(\mu_{2\alpha} + \mu)}\psi_{2\alpha}.$$

Now

(4.16) $$\mu^{-1} = \int_0^\infty e^{-\mu s}\, ds \quad \text{if} \quad Re\mu > 0.$$

We use (4.16) with $\mu$ replaced by $\mu + \mu_{2\alpha}$, insert it into (4.15), and obtain

(4.17)
$$
\begin{aligned}
q_\lambda &= \sum_\alpha \frac{2^{n-2|\alpha|}}{\alpha!}\left(\int_0^\infty e^{-(\mu_{2\alpha}+\mu)s}\, ds\right)\psi_{2\alpha} \\
&= 2^n \int_0^\infty e^{-\mu s}\left\{\sum_\alpha \frac{2^{-2|\alpha|}}{\alpha!}e^{-\mu_{2\alpha}}\psi_{2\alpha}\right\}ds \\
&= \int_0^\infty e^{-\mu s} G(\xi, s)\, ds,
\end{aligned}
$$

where

(4.18) $$G(\xi, s) = \prod_{j=1}^{2n} g\left(\beta_j \xi_j, \frac{1}{2}a_j|\xi_0|s\right),$$

with

(4.19) $$g(x, s) = 2^{\frac{1}{2}}\sum_{k=0}^\infty \frac{2^{-2k}}{k!}e^{-(4k+1)s}\psi_{2k}(x).$$

To evaluate $g$ we continue our formal reasoning. From (4.4),

(4.20) $$\psi_k(x) = e^{\frac{1}{2}x^2}\left(-\frac{d}{dx}\right)^k e^{-x^2},$$

so

(4.21) $$\sum_{k=0}^\infty \frac{1}{k!}\left(\frac{1}{2}y\right)^{2k}\psi_{2k}(x) = e^{\frac{1}{2}x^2}\left[\sum_{k=0}^\infty \frac{1}{k!}\left(\frac{y}{2}\frac{d}{dx}\right)^{2k}e^{-x^2}\right].$$

The expression in brackets is evaluated by taking the Fourier transform in $x$, summing, and taking the inverse Fourier transform. Replacing $y$ by $e^{-2s}$

gives

(4.22)            $$g(x, s) = [\cosh(2s)]^{-\frac{1}{2}} \exp\left(-\frac{1}{2}x^2 \tanh(2s)\right).$$

Thus in (4.18),
(4.23)

$$G(\xi, s) = \prod_{j=1}^{2n} \cosh(|\xi_0|a_j s)^{-\frac{1}{2}} \exp\left[-\sum_{j=1}^{2n} \xi_j^2 \tanh(|\xi_0|a_j s)(|\xi_0|a_j)^{-1}\right].$$

Taking the limit as $\xi_0 \to 0$ gives

(4.24)            $$\lim_{\xi_0 \to 0} G(\xi_0, \xi_1, \ldots, \xi_{2n}, s) = \exp\left(-\sum_{j=1}^{2n} \xi_j^2 s\right).$$

A direct calculation shows that

(4.25)                            $$\frac{\partial}{\partial s}G = -LG$$

so a formal integration by parts in (4.17) gives

$$(L + \mu)q_\lambda = \int_0^\infty e^{-\mu s}\left[\mu - \frac{\partial}{\partial s}\right]G(\cdot, s)ds$$

(4.26)
$$= -\int_0^\infty \frac{\partial}{\partial s}\{e^{-\mu s}G(\cdot, s)\}ds$$

$$= G(\cdot, 0) = 1.$$

The formal argument is complete. We want now to justify the integral (4.17) and the calculation (4.26). Examination of the hyperbolic cosine terms in (4.23) shows

(4.27)            $$|G(\xi, s)| \leq 2^n \exp\left(-\frac{1}{2}\sum_{j=1}^{2n} |\xi_0|a_j s\right),$$

for $s \geq 0$. Therefore the integral (4.17) is rapidly convergent and (4.26) is valid provided

(4.28)                      $$-Re\,\mu < \frac{1}{2}\sum_{j=1}^{2n} |\xi_0|a_j.$$

We seek to extend $q_\lambda$ from the region (4.28) by analytic continuation in $\mu$. Indeed we start from the smaller region

(4.29)                            $$|\arg \mu| < \frac{1}{2}\pi.$$

Now $G(\xi, \cdot)$ is holomorphic in $s$ on $\mathbf{C}\backslash i\mathbf{R}$ and converges rapidly to zero as $s$ tends to $\infty$ along any ray in the right half plane. Therefore the integral

$$(4.30) \qquad \int_0^\infty e^{-\omega\mu s} G(\xi, \omega s)\, d(\omega s)$$

is rapidly convergent for any $\omega \in \mathbf{C}\backslash(0)$ such that

$$(4.31) \qquad |\arg \omega| < \frac{1}{2}\pi, \quad |\arg \omega + \arg \mu| < \frac{1}{2}\pi.$$

When $\omega > 0$ we obtain (4.17). Starting from $\mu$ such that

$$\frac{1}{2}\pi - \varepsilon < \arg \mu < \frac{1}{2}\pi$$

for some $\varepsilon \in (0, \frac{1}{2}\pi)$, we may deform the contour of integration in (4.17) by moving $\omega$ in (4.30) clockwise along the unit circle to the point where $\arg \omega = -(\frac{1}{2}\pi - \varepsilon)$. On this new contour the integral converges in the range

$$-\varepsilon < \arg \mu < \pi - \varepsilon.$$

Similarly, starting from

$$-\frac{1}{2}\pi < \arg \mu < -\frac{1}{2}\pi + \varepsilon$$

we may move to $\arg \omega = \frac{1}{2}\pi - \varepsilon$ and obtain an integral convergent in the range

$$-\pi + \varepsilon < \arg \mu < \varepsilon.$$

Thus (4.17) can be continued in $\mu$ to the entire region $C\backslash\mathbf{R}_-$.

If $2n = d$ we should be able to extend (4.17) to the entire region

$$(4.32) \qquad -\mu \neq \mu_\alpha, \quad \text{all} \quad \alpha \in 2\mathbf{Z}_+^{2n},$$

i.e.

$$(4.33) \qquad \lambda \neq \pm \sum_{j=1}^n (2\alpha_j + 1)a_j, \quad \alpha \in \mathbf{Z}_+^n.$$

where $\mu_\alpha$ is given by (4.9). This can be done by repeated integrations by parts in (4.17). The procedure uses an identity for $G$ which we derive formally by returning to the Hermite functions (4.4). From (4.4) it follows that

$$(4.34) \qquad \psi_{k+1}(x) = \left(x - \frac{d}{dx}\right)\psi_k(x).$$

The formal series (4.19) for the function $g$ then suggests the identity

(4.35) $$\left(\frac{\partial}{\partial s} + 1\right)g(x,s) = -e^{-4s}\left(\frac{\partial}{\partial x} - x\right)^2 g(x,s).$$

This identity is directly verifiable from (4.22) and the corresponding identity for $G$ is

(4.36) $$\left(\frac{\partial}{\partial s} + \gamma\right)G = \sum_{j=1}^{2n} e^{-2a_j|\xi_0|s} M_j G$$

where

(4.37) $$\gamma = \gamma(\xi_0) = \frac{1}{2}\sum_{j=1}^{2n} a_j|\xi_0|,$$

and

(4.38) $$M_j = -\left(\frac{1}{2}a_j|\xi_0|\frac{\partial}{\partial \xi_j} - \xi_j\right)^2.$$

For $s \geq 0$,

(4.39) $$|M_j G| \leq C_j e^{-\gamma s}.$$

Therefore in the region $-\mu < \gamma$ we may integrate by parts in (4.17) to obtain

$$q_\lambda = \int_0^\infty e^{-\mu s} G(\xi, s)ds$$

(4.40) $$= -(\mu + \gamma)^{-1}\int_0^\infty \frac{d}{ds}(e^{-\mu s - \gamma s})e^{\gamma s} G(\xi, s)da$$

$$= (\mu + \gamma)^{-1}\left[G(\xi, 0) + \int_0^\infty e^{-\mu s}\left(\frac{\partial}{\partial s} + \gamma\right)G(\xi, s)ds\right]$$

or

(4.41) $$q_\lambda = (\mu + \gamma)^{-1}\left[1 + \sum_{j=1}^{2n}\int_0^\infty e^{-\mu s - 2a_j|\xi_0|s} M_j G\, ds\right].$$

In view of (4.39) the integral in (4.41) converges in the region

(4.42) $$-Re\,\mu < \gamma + 2\min_j a_j|\xi_0|.$$

Therefore (4.41) gives the analytic continuation of $q_\lambda$ to the region (4.42) except for a pole at $\mu + \gamma = 0$. This procedure may be repeated indefinitely.

At the second step we apply the procedure of (4.40) to each integral in (4.41) with $\mu$ replaced by $\mu + 2a_j|\xi_0|$. The resulting integrals involve terms $M_j M_k G$ which satisfy estimates lilke (4.39). This extends $q_\lambda$ to the region

$$(4.43) \qquad -Re\,\mu < \gamma + 4\min_j a_j|\xi_0|$$

except for poles at $\mu + \gamma = 0$ and $\mu + \gamma + 2a_j|\xi_0| = 0$, $j = 1,\ldots,2n$. Iterating, we obtain the continuation to the entire region (4.33). We note that $M_j M_k = M_k M_j$.

To describe the general form of the resulting expression for $q_\lambda$ we introduce some notation. If $J = (j_1, j_2, \ldots, j_m)$ is an ordered subset of $\{1, 2, \ldots, 2n\}^m$, set $|J| = m$ and

$$(4.44) \qquad \mu_J = \mu_J(\xi_0) = \prod_{k=0}^{m}\left(\mu + \gamma + \sum_{l \le k} 2a_{j_l}|\xi_0|\right).$$

(The factor with $k = 0$ is taken to be $\mu + \gamma$.) Let $D_J$ denote the operator

$$(4.45) \qquad D_J = \mu_J^{-1}\exp\left(-2\sum_{l=1}^{m}a_{j_l}|\xi_0|\right)M_{j_1}M_{j_2}\ldots M_{j_m}.$$

Then at the $(m+1)-st$ step of the procedure described above we obtain

$$(4.46) \qquad q_\lambda = \sum_{|J| \le m} D_J G(\xi,0) + \sum_{|J|=m}\int_0^\infty e^{-\mu s}\left(\frac{\partial}{\partial s} + \gamma\right)D_J G(\xi, s)ds.$$

This identity is valid in the original region $-Re\,\mu < \gamma$ and extends to

$$(4.47) \qquad -Re\,\mu < \gamma + 2(m+1)\min_j a_j|\xi_0|$$

except for the zeros of the functions $\mu_J$.

As in §1 we define the *singular set* of the operator (3.1) by

$$(4.48) \qquad \Lambda = \mathbf{R} \quad \text{if} \quad n = 0;$$

$$(4.49) \qquad \Lambda = \left\{\pm\sum_{j=1}^{n}(2\alpha_j + 1)a_j : \alpha \in Z_+^n\right\} \quad \text{if} \quad 2n = d;$$

$$(4.50) \qquad \Lambda = \left\{\lambda \in \mathbf{R}, |\lambda| \ge \sum_{j=1}^{n}a_j\right\} \quad \text{if} \quad 2n < d.$$

(4.51) THEOREM: *If $\lambda \notin \Lambda$, then the operator $P_\lambda$ given by (4.1) has a two-sided pseudodifferential inverse $Q_\lambda$ with symbol*

$$(4.52) \qquad \sigma(P_\lambda^{-1})(x,\xi) = q_\lambda(\sigma_0(x,\xi),\ldots,\sigma_d(x,\xi))$$

*where $\sigma_j$, given by (3.20), is the symbol of $i^{-1}X_j$. The function $q_\lambda$ is defined
by*

(4.53) $$q_\lambda(\xi) = \int_0^\infty e^{-\mu s} G(\xi, s) ds \quad if \quad -Re\,\mu < \gamma,$$

*where*

(4.54) $$\mu = \mu(\xi) = \lambda \xi_0 + \sum_{j=2n+1}^d \xi_j^2$$

*and*
(4.55)
$$G(\xi, s) = \left[ \prod_{j=1}^{2n} \cosh(a_j |\xi_0| s) \right]^{-\frac{1}{2}} \exp\left( -\sum_{j=1}^{2n} \xi_j^2 \tanh(a_j |\xi_0| s)(a_j |\xi_0|)^{-1} \right),$$

*if $n \neq 0$, or*

(4.56) $$q_\lambda(\xi) = \left( \lambda \xi_0 + \sum_{j=1}^d \xi_j^2 \right)^{-1} \quad if \quad n = 0.$$

*If $0 < 2n < d$ then on the remaining region, $\{\mathrm{Im}\,\mu \neq 0\}$, $q_\lambda$ is obtained
from (4.53) by analytic continuation, changing the contour of integration –
see (4.30) and consequent discussion.*

*If $2n = d$, then (4.46) continues $q_\lambda$ analytically to the region (4.47) for
all $m = 1, 2, 3, \ldots$.*

Proof. We set aside the case $n = 0$ (which can be easily handled on its
own merits or by letting $a_j \to 0$ in the formulas above). $q_\lambda$ is smooth on
$\mathbf{R}^{d+1} \backslash 0$ and has the mixed homogeneity,

(4.57) $$q_\lambda(t^2 \xi_0, t\xi_1, \ldots, t\xi_d) = t^{-2} q_\lambda(\xi_0, \xi_1, \ldots, \xi_d), \quad t > 0.$$

Therefore

(4.58) $$|D_\xi^\alpha q_\lambda(\xi)| \leq C_\alpha \langle \xi \rangle^{-2-\alpha_0-|\alpha|},$$

where $|\alpha| = \sum_{j=0}^d \alpha_j$ and

(4.59) $$\langle \xi \rangle = \left( |\xi_0| + \sum_{j=1}^d \xi_j^2 \right)^{\frac{1}{2}}.$$

We note that $q_\lambda$ is holomorphic in $\lambda$ when $\lambda$ is not in the singular set $\Lambda$ and satisfies the differential equation (3.32) for $|Re\ \lambda| < \sum_{j=1}^{n} a_j$ – see (4.26). Therefore it satisfies (3.32) for all $\lambda \notin \Lambda$.

Next define the operator $Q_\lambda$ by (3.6), i.e.

$$(4.60) \qquad Q_\lambda u(x) = (2\pi)^{-d-1} \int_{\mathbf{R}^{d+1}} e^{i\langle x,\xi\rangle} \sigma(Q_\lambda)(x,\xi)\hat{u}(\xi)d\xi,$$

$u \in \mathcal{S}$, where

$$(4.61) \qquad \sigma(Q_\lambda)(x,\xi) = q_\lambda(\sigma(x,\xi)),$$

with $\sigma$ given by (3.20) and $(b_{jk})$ in normal form, i.e given by (1.26) and (1.27). The function $q_\lambda(\sigma(x,\xi))$ does not belong to $S^\infty$; however it is in $C^\infty(\mathbf{R}^{d+1} \times (\mathbf{R}^{d+1}\backslash 0))$, satisfies the estimate (3.8) for $\langle\xi\rangle \geq 1$ and is integrable near $\xi = 0$. Therefore (4.60) is well defined. Furthermore

$$(4.62) \qquad D_x^\alpha q_\lambda(\sigma(x,\xi)) = q_\lambda^{(\alpha)}(\sigma(x,\xi)),$$

where $q_\lambda^{(\alpha)}$ is homogeneous of degree $-2 + \alpha_0 + |\alpha|$; thus $D_x^\alpha q_\lambda$ is also integrable at $\xi = 0$ and we see that

$$(4.63) \qquad Q_\lambda : \mathcal{S} \to C^\infty.$$

Despite the lack of smoothness of $q_\lambda$ at $\xi = 0$, the integrability at $\xi = 0$ permits the use of the dominated convergence theorem which justifies the calculations in the proof of Proposition 3.10. Thus $Q_\lambda P_\lambda$ and $P_\lambda Q_\lambda$ have symbols given by

$$(4.64) \qquad \sum_\alpha \frac{1}{\alpha!}\partial_\xi^\alpha \sigma(Q_\lambda)D_x^\alpha p_\lambda,$$

$$(4.65) \qquad \sum_\alpha \frac{1}{\alpha!}\partial_\xi^\alpha p_\lambda D_x^\alpha \sigma(Q_\lambda),$$

respectively. According to (3.32) and the discussion following (4.59) these symbols $\equiv 1$ at $x = 0$, and since $P_\lambda$ and $Q_\lambda$ are both left-invariant with respect to the group law (3.18), we also have that (4.64) and (4.65) are $\equiv 1$ for all $x$. Thus we have shown that

$$(4.66) \qquad Q_\lambda P_\lambda = P_\lambda Q_\lambda = Id,$$

when $\lambda \notin \Lambda$. This proves Theorem 4.51.

(4.67) REMARKS: Our procedure for calculating the symbol of the inverse of $P_\lambda$ implicitly introduces the Laplace transform via the identity (4.16).

In effect we are computing the symbol of the inverse of the heat operator $\frac{\partial}{\partial t} + P_\lambda$, or equivalently the symbol of $\exp(-tP_\lambda)$, to obtain

$$P_\lambda^{-1} = \int_0^\infty e^{-tP_\lambda} \, dt.$$

Hulanicki [1] calculated the kernel of the inverse of $\frac{\partial}{\partial t} + P_\lambda$ in the special case of the Kohn laplacian on the Heisenberg group in essentially the same way as done here. Gaveau [1] independently computed the kernel in the same special case by probabilistic methods.

## §5 Inverting $P_\lambda$ in Normal Form: Kernels

A pseudodifferential operator $Q$ with symbol $q$ is, at least formally, an integral operator whose kernel is the partial inverse Fourier transform of the symbol:

$$Qu(x) = (2\pi)^{-m} \int_{\mathbf{R}^m} e^{i\langle x,\xi\rangle} q(x,\xi) \int_{\mathbf{R}^m} e^{-i\langle y,\xi\rangle} u(y) \, dy d\xi$$

(5.1)
$$= \int_{\mathbf{R}^m} \left[ (2\pi)^{-m} \int_{\mathbf{R}^m} e^{i\langle x-y,\xi\rangle} q(x,\xi) \, d\xi \right] u(y) dy$$

$$= \int_{\mathbf{R}^m} K(x, x-y) u(y) dy.$$

On the other hand, an operator $Q$ which is left invariant with respect to the group structure (3.18) ought to be given by convolution:

(5.2)
$$Qu(x) = (k * u)(x) = \int k(y) u(xy^{-1}) dy$$

$$= \int k(y^{-1}x) u(y) dy.$$

(The Lebesgue measure $dx$ is a Haar measure for the group.)

The group (3.18) admits a 1-parameter automorphism group, the dilations

(5.3)
$$\delta_t x = (t^2 x_0, t x_1, \ldots, t x_d), \quad t > 0.$$

The operator $P$ of (4.1) is homogeneous of degree 2 with respect to these dilations, that is, if we let

(5.4) $$V_t u = u \circ \delta_t, \quad u \in C_c^\infty(\mathbf{R}^{d+1}),$$

then

(5.5) $$V_t^{-1} P V_t = t^2 P, \quad t > 0.$$

If the inverse $Q$ is indeed given by a convolution (5.2) with unique kernel $k$, then (5.5) gives

(5.6) $$k \circ \delta_t = t^{-d} k.$$

In fact $Q$ must be homogeneous of degree -2, so

(5.7) $$k * (u \circ \delta_t) = t^{-2}(k * u) \circ \delta_t.$$

But

(5.8)
$$\begin{aligned}
k * (u \circ \delta_t)(x) &= \int k(y^{-1}x)u(\delta_t y)\,dy \\
&= t^{-d-2} \int k(\delta_t^{-1}(y)^{-1}x)u(y)\,dy \\
&= t^{-d-2}[(k \circ \delta_t^{-1}) * u] \circ \delta_t(x),
\end{aligned}$$

and (5.7), (5.8) give (5.6).

The following justifies the preceding formal argument.

(5.9) THEOREM: *Suppose the operator $P_\lambda$ of (4.1) is invertible and let $q_\lambda$ be the function of Theorem 4.51, which determines the symbol of the inverse $Q_\lambda$. Let $k_\lambda$ be the inverse Fourier transform of $q_\lambda$ as tempered distribution. Then $k_\lambda$ is locally integrable, $k_\lambda$ is $C^\infty$ except at the origin,*

(5.10) $$k_\lambda \circ \delta_t = t^{-d} k_\lambda, \quad t > 0,$$

*and*

(5.11) $$Q_\lambda u(x) = \int k_\lambda(y^{-1}x)u(y)\,dy, \quad u \in C_c^\infty(\mathbf{R}^{d+1}).$$

Proof. Any distribution derivative of $q_\lambda$ of sufficiently high order is the sum of an integrable function (the corresponding derivative of $q_\lambda$ truncated near the origin) and a distribution with compact support. For the inverse Fourier transform $k_\lambda$ this means $x^\alpha k_\lambda(x)$ is continuous for all large $|\alpha|$. Thus $k_\lambda$ is continuous on $\mathbf{R}^{d+1}\backslash\{0\}$. The distribution derivative $D^\alpha k_\lambda$ is the inverse Fourier transform of $\xi^\alpha q_\lambda(\xi)$ and is continuous on $\mathbf{R}^{d+1}\backslash\{0\}$ by the same argument. (4.52) means

$$(5.12) \qquad\qquad q_\lambda \circ \delta_t = t^{-2} q_\lambda$$

as distributions, and a straightforward calculation shows that (5.10) holds as distributions. In particular the *function* $k_\lambda$ on $\mathbf{R}^{d+1}\backslash\{0\}$ is homogeneous of degree $-d$ with respect to the dilations (5.3) and is therefore integrable at 0. The corresponding distribution differs from the *distribution* $k_\lambda$ by a distribution which is supported at 0 and has the same homogeneity property (5.10). This implies that the difference vanishes, and the distribution $k_\lambda$ is a function.

To prove (5.11) at $x = 0$ we carry out (5.1) with a regularization:

$$
\begin{aligned}
(5.13) \qquad Q_\lambda u(0) &= \lim_{\varepsilon \searrow 0} (2\pi)^{-d-1} \int q_\lambda(\xi) e^{-\varepsilon|\xi|^2} \int e^{-i\langle y,\xi\rangle} u(y)\,dy\,d\xi \\
&= \lim_{\varepsilon \searrow 0} \int \left\{ (2\pi)^{-d-1} \int e^{-i\langle y,\xi\rangle - \varepsilon|\xi|^2} q_\lambda(\xi)\,d\xi \right\} u(y)\,dy.
\end{aligned}
$$

The expression in braces converges as distribution to $k_\lambda(-y) = k_\lambda(y^{-1})$, so (5.11) is true at $x = 0$. In general let $\psi(y) = xy$ and use translation invariance of $Q_\lambda$:

$$
\begin{aligned}
(5.14) \qquad Q_\lambda u(x) &= Q_\lambda(u \circ \psi)(0) = \int k_\lambda(y^{-1}) u(xy)\,dy \\
&= \int k_\lambda(y^{-1}x) u(y)\,dy.
\end{aligned}
$$

This proves Theorem 5.9.

We seek now to compute the kernel $k_\lambda$ using (4.52), (4.53) and (4.55). When $\xi_0 \neq 0$ we change variables:

$$(5.15) \qquad q_\lambda(\xi) = \int_0^\infty |\xi_0|^{-1} e^{-\mu s} G(\xi, |\xi_0|^{-1} s)\,ds,$$

where now

$$(5.16) \qquad \mu = \mu(\xi) = \lambda \operatorname{sgn} \xi_0 + \left( \sum_{j=2n+1}^{d} \xi_j^2 \right) |\xi_0|^{-1}.$$

We begin with the inverse Fourier transform in the variables $\xi_1, \ldots, \xi_d$. The integrand of (5.15) is a multiple of

(5.17)
$$\exp\left(-\frac{1}{2}\sum_{j=1}^{d}\beta_j\xi_j^2\right),$$

$$\beta_j = 2a_j^{-1}|\xi_0|^{-1}\tanh(a_js), \quad 1 \le j \le 2n,$$

$$= 2s|\xi_0|^{-1}, \quad 2n < j \le d.$$

The inverse Fourier transform of (5.17) in the variables $\xi_1, \ldots, \xi_d$ is

(5.18)
$$\left[\prod_{j=1}^{d}(2\pi\beta_j)^{-\frac{1}{2}}\right]\exp\left(-\frac{1}{2}\sum_{j=1}^{d}\beta_j^{-1}x_j^2\right).$$

Thus the inverse Fourier transform of the integrand in (5.15) with respect to the variables $\xi_1, \ldots, \xi_d$ is

(5.19)
$$s^{-\frac{1}{2}d}A(s)|\xi_0|^{-1+\frac{1}{2}d}\exp[-s\lambda\,\text{sgn}\,\xi_0 - s^{-1}\gamma(x',s)|\xi_0|],$$

where $x' = (x_1, \ldots, x_d)$ and

(5.20)
$$A(s) = (4\pi)^{-\frac{1}{2}d}\prod_{j=1}^{n}\text{cosech}(a_js),$$

(5.21)
$$\gamma(x',s) = \frac{1}{4}\sum_{j=1}^{2n}a_jsx_j^2\coth(a_js) + \frac{1}{4}\sum_{j=2n+1}^{d}x_j^2.$$

The inverse Fourier transform of the function

(5.22)
$$|\xi_0|^{r-1}e^{-b\,sgn\xi_0-\delta|\xi_0|}, \quad r, \delta > 0,$$

is

(5.23)
$$(2\pi)^{-1}\Gamma(r)[e^{-b}(\delta - ix_0)^{-r} + e^{b}(\delta + ix_0)^{-r}],$$

where we take the principal branch of the power. Thus the inverse Fourier transform of the integrand in (5.15) is

(5.24)
$$(2\pi)^{-1}\Gamma\left(\frac{1}{2}d\right)A(s)\{e^{-\lambda s}[\gamma(x',s) - ix_0s]^{-\frac{1}{2}d}$$
$$+ e^{\lambda s}[\gamma(x',s) + ix_0s]^{-\frac{1}{2}d}\}.$$

Both $A$ and $\gamma$ are even functions of $s$, so we have

(5.25)
$$k_\lambda(x) = (2\pi)^{-1}\Gamma\left(\frac{1}{2}d\right)\int_{-\infty}^{\infty}A(s)e^{-\lambda s}[\gamma(x',s) - ix_0s]^{-\frac{1}{2}d}ds.$$

The computation leading to (5.25) is clearly valid when $x' \neq 0$ and

(5.26)
$$|\operatorname{Re}\lambda| < \sum_{j=1}^{n} a_j.$$

When $x_0 \neq 0$ and $x' = 0$, then $\gamma(x', s) - ix_0 s = -ix_0 s$ and the integrand is not integrable at $s = 0$. To regularize the integral we shall deform its path of integration. To do this we need a better understanding of $\gamma - ix_0 s$. Set $z = u + iv$. Then

(5.27)
$$z \coth z = \frac{u\sinh(2u) + v\sin(2v)}{\cosh(2u) - \cos(2v)} + i\frac{v\sinh(2u) - u\sin(2v)}{\cosh(2u) - \cos(2v)}.$$

Now

$$\cosh(2u) - \cos(2v) = 0 \Longrightarrow u = 0,$$

and, when $u = 0$

$$iv\coth(iv) = v\frac{\cos v}{\sin v}.$$

Consequently, $z \coth z$ is well defined as long as $z \neq \pm ik\pi$, $k = 1, 2, 3\dots$. This is, of course, elementary. On the other hand, (5.27) also implies that

$$\operatorname{Re}(z \coth z) > 0 \quad \text{if } -\frac{\pi}{2} < \operatorname{Im} z < \frac{\pi}{2},$$

Applying (5.28) to $\gamma - ix_0 s$ we find

$$\operatorname{Re}[\gamma(x', s) - ix_0 s]$$

(5.29)
$$= \sum_{j=1}^{2n} \frac{x_j^2}{4} \operatorname{Re}[a_j s \coth(a_j s)] + \sum_{j=2n+1}^{d} \frac{x_j^2}{4} + x_0 \operatorname{Im} s > 0$$

as long as

(5.30)
$$(x_0 \operatorname{Im} s, x') \neq 0,$$

(5.31)
$$x_0 \operatorname{Im} s \geq 0,$$

and

(5.32)
$$|\operatorname{Im} s| < \frac{\pi}{2a_j}, \quad j = 1, \dots, n.$$

*Therefore the path of integration in (5.25) can be deformed to* $(-\infty + i\varepsilon \ \operatorname{sgn} x_0, \infty + i\varepsilon \ \operatorname{sgn} x_0)$, $0 < \varepsilon < \min_j \frac{1}{2} a_j^{-1}\pi$, *i.e.*

(5.33)
$$k_\lambda(x) = (2\pi)^{-1}\Gamma\left(\frac{d}{2}\right) \int_{-\infty + i\varepsilon \ \operatorname{sgn} x_0}^{\infty + i\varepsilon \ \operatorname{sgn} x_0} A(s)e^{-\lambda s}[\gamma(x', s) - ix_0 s]^{-d/2} ds.$$

*For sufficiently small $\varepsilon > 0$ (5.33) agrees with (5.25) when $x' \neq 0$ – simply by moving the path of integration to $(-\infty, \infty)$. In particular it defines $k_\lambda(x', 0)$ as long as $x' \neq 0$. Moreover it yields the necessary regularization when $x' = 0$ and $x_0 \neq 0$.*

Set $s = u + iv$. Then an elementary calculation yields the following formula:

$$
\begin{aligned}
|\gamma(x', s) - ix_0 s|^2 = (u^2 + v^2) &\left( \left[ \sum_{j=1}^{2n} \frac{x_j^2}{4} \frac{a_j \sinh(2a_j u)}{\cosh(2a_j u) - \cos(2a_j v)} \right.\right. \\
&\left. + \sum_{j=2n+1}^{d} \frac{x_j^2}{4} \frac{u}{u^2 + v^2} \right]^2 \\
&\left. + \left[ x_0 + \sum_{j=1}^{2n} \frac{x_j^2}{4} \frac{a_j \sin(2a_j v)}{\cosh(2a_j u) - \cos(2a_j v)} + \sum_{j=2n+1}^{d} \frac{x_j^2}{4} \frac{v}{u^2 + v^2} \right]^2 \right).
\end{aligned}
$$

(5.34)

For the derivation of (5.34) it is useful to assume first that $a_j > 0, j = 1, \dots, d$, then, after computing $|\gamma(x', s) - ix_0 s|^2$, let $a_j \to 0$, $j = 2n + 1, \dots, d$. This formula is remarkably simple and elegant. Furthermore (5.29) - (5.32) and (5.34) yield

(5.35) PROPOSITION: *Assume $\gamma(x', s) - ix_0 s$ is well defined, that is $a_j s \neq \pm ik\pi$, $j = 1, \dots, n$ and $k = 0, 1, 2, \dots$. Then*

  (i) $x \neq 0$, $\operatorname{Re} s \neq 0 \Longrightarrow \gamma(x', s) - ix_0 s \neq 0$
  (ii) $(x_0 \operatorname{Im} s, x') \neq 0, x_0 \operatorname{Im} s \geq 0$, $|\operatorname{Im} s| < \min_j \frac{1}{2} a_j^{-1} \pi, \Longrightarrow \operatorname{Re}(\gamma(x', s) - ix_0 s) > 0$
  (iii) $\operatorname{Re} s = 0 \Longrightarrow$ *for every given $v$ and $v'$ one can always find an $x_0$ with $x_0 v \leq 0$ such that $\gamma(x', s) - ix_0 s = 0$.*

To continue (5.25) from the region (5.26) we move the contour of integration. The poles of $A(s)$, and if $x \neq 0$, the poles and zeroes of $\lambda(x', s) - ix_0 s$ lie on the imaginary axis. Therefore the contour may be changed so as to coincide near $s = \infty$ with any given line through the origin, provided that at each stage of the deformation

$$
(5.36) \qquad |\operatorname{Re}(\lambda s)| < \sum_{j=1}^{n} a_j |\operatorname{Re} s| \quad \text{as} \quad s \to \infty.
$$

Suppose we deform the integration contour into a given line $\mathbf{R}\omega$, $\omega = e^{i\theta}$, $-\frac{\pi}{2} < \theta < \frac{\pi}{2}$. Let $\lambda = |\lambda| e^{i\alpha}$. Then (5.36) is equivalent to

$$
|\lambda \cos(\alpha + \theta)| < \sum_{j=1}^{n} a_j \cos \theta.
$$

Thus if $\lambda$ is in a sufficiently small neighborhood of the origin we are justified to deform the contour $(-\infty, \infty)$ into $\mathbf{R}\omega$, asymptotically. Once this change is made we may continue analytically in $\lambda$ throughout the region where

$$(5.37) \qquad |\operatorname{Re}(\lambda\omega)| < \sum_{j=1}^{n} a_j |\operatorname{Re}\omega|.$$

In particular, this includes the line $i\mathbf{R}\bar{\omega}$. Consequently, for every $\lambda \in \mathbf{C}\backslash\mathbf{R}$ there is a contour $I(\lambda)$, so that (5.25) gives $k_\lambda(x)$ if we replace the path $(-\infty, \infty)$ by $I(\lambda)$.

(5.38) THEOREM: *Suppose $\lambda \in C$ satisfies either*

$$(5.39) \qquad |\operatorname{Re}\lambda| < \sum_{j=1}^{n} a_j$$

*or*

$$(5.40) \qquad \lambda \notin \mathbf{R}.$$

*Then there is a contour $I = I(\operatorname{sgn} x_0, \lambda)$ such that the convolution kernel $k_\lambda$ giving the inverse of $P_\lambda$ is*

$$(5.41) \qquad k_\lambda(x) = \frac{1}{2\pi}\Gamma\left(\frac{1}{2}d\right)\int_I A(s)e^{-\lambda s}[\gamma(x', s) - ix_0 s]^{-\frac{1}{2}d}\,ds$$

*where $A$ and $\gamma$ are given by (5.20) and (5.21).*

(5.42) REMARK: Both the derivation and the formula remain valid if $n = 0$, and indeed if the $a_j$ all converge to zero then (5.20) and (5.21) become

$$A(s) = (4\pi)^{-\frac{1}{2}d}, \quad \gamma(x', s) = \frac{1}{4}\sum_{j=1}^{d} x_j^2.$$

In this case (5.41) can be computed as

$$k_\lambda(x) = \frac{1}{2}(4\pi|x_0|)^{-\frac{1}{2}d}[1 + \operatorname{sgn}(x_0 \operatorname{Im}\lambda)]$$

$$(5.43)$$

$$[-i\lambda \operatorname{sgn}(\operatorname{Im}\lambda)]^{\frac{1}{2}d-1}\exp\left(\frac{1}{4}i\lambda\sum_{j=1}^{d} x_0^{-1}x_j^2\right).$$

First compute for $\operatorname{Re}\lambda = 0$, then continue analytically. For another derivation see Remark 5.6.1.

When $2n < d$ we have computed the convolution kernel for all $\lambda$ such that $P_\lambda$ of (4.1) is invertible. When $2n = d$ there is only a discrete set of inadmissible values of $\lambda$ on the real axis, and for the real axis our calculation is valid, so far, only in the interval containing the origin. To extend to other admissible real $\lambda$ we use the procedure developed in Chapter 4 to continue the symbol analytically. Let us put (5.33) in the following form:

$$(5.44) \qquad k_\lambda(x) = \int_{-\infty + i\varepsilon\, \mathrm{sgn}\, x_0}^{\infty + i\varepsilon\, \mathrm{sgn}\, x_0} e^{-\lambda s} H(x, s)\, ds,$$

where $\lambda$ is in the region (5.26). Note that changing the sign of the variable $x_0$ is equivalent to changing the sign of $\lambda$ in (5.25), so we may assume in what follows that

$$(5.45) \qquad \mathrm{Re}\, \lambda \leq 0.$$

The integral

$$(5.46) \qquad k_\lambda^-(x) = \int_{-\infty + i\varepsilon\, \mathrm{sgn}\, x_0}^{i\varepsilon\, \mathrm{sgn}\, x_0} e^{-\lambda s} H(x, s)\, ds$$

is holomorphic in $\lambda$ in the entire region (5.45), so we need only extend

$$(5.47) \qquad k_\lambda^+(x) = \int_{i\varepsilon\, \mathrm{sgn}\, x_0}^{\infty + i\varepsilon\, \mathrm{sgn}\, x_0} e^{-\lambda s} H(x, s)\, ds.$$

We use the (formal) identity

$$k_\lambda^+(x) = \int_{i\varepsilon\, \mathrm{sgn}\, x_0}^{\infty + i\varepsilon\, \mathrm{sgn}\, x_0} e^{-\lambda s} H(x, s)\, ds$$

$$(5.48) \qquad = \frac{1}{\lambda + \alpha} H(x, i\varepsilon\, \mathrm{sgn}\, x_0)$$

$$+ \frac{1}{\lambda + \alpha} \int_{i\varepsilon\, \mathrm{sgn}\, x_0}^{\infty + i\varepsilon\, \mathrm{sgn}\, x_0} e^{-\lambda s} \left[ \frac{\partial}{\partial s} + \alpha \right] H(x, s)\, ds,$$

and take

$$(5.49) \qquad \alpha = \sum_{j=1}^{n} a_j.$$

Recall that

$$\hat{H}(\xi, s) = \frac{1}{|\xi_0|} G\left( \xi, \frac{s}{|\xi_0|} \right),$$

where $\hat{}$ denotes the Fourier transform in the $x$ variables. According to (4.36)

$$\left(\frac{\partial}{\partial t} + \sum_{j=1}^{n} a_j |\xi_0|\right) G(\xi, t) = \sum_{j=1}^{2n} e^{-2a_j |\xi_0| t} M_j G(\xi, t).$$

We set $t = s/|\xi_0|$ and divide both sides by $|\xi_0|^2$. Then

$$\left(\frac{\partial}{\partial s} + \sum_{j=1}^{n} a_j\right) \frac{1}{|\xi_0|} G\left(\xi, \frac{s}{|\xi_0|}\right) = \sum_{j=1}^{2n} e^{-2a_j s} \frac{M_j}{|\xi_0|} \left[\frac{1}{|\xi_0|} G\left(\xi, \frac{s}{|\xi_0|}\right)\right].$$

We set

$$(5.50) \qquad \check{M}_j = i\left(\frac{\partial}{\partial x_j} + \frac{1}{2} i a_j x_j \frac{\partial}{\partial x_0}\right) \mathcal{L} \left(\frac{\partial}{\partial x_j} + \frac{1}{2} i a_j x_j \frac{\partial}{\partial x_0}\right),$$

where $\mathcal{L}$ is the integration operator in $x_0$:

$$(5.51) \qquad \mathcal{L} f(x_0, x') = \int_{\infty}^{x_0} f(t, x') dt.$$

Then a direct calculation proves the following identity

$$(5.52) \qquad \left(\frac{\partial}{\partial s} + \alpha\right) H = \sum_{j=1}^{2n} e^{-2a_j s} \check{M}_j H.$$

We note that (5.52) is the inverse Fourier transform of (4.36). Thus in the region $-\alpha < \operatorname{Re}\lambda \leq 0$ we can write (5.48) in the following form

$$k_\lambda^+(x) = \frac{1}{\lambda + \alpha} H(x, i\varepsilon \operatorname{sgn} x_0)$$

(5.53)

$$+ \frac{1}{\lambda + \alpha} \int_{i\varepsilon \operatorname{sgn} x_0}^{\infty + i\varepsilon \operatorname{sgn} x_0} e^{-\lambda s} \sum_{j=1}^{2n} e^{-2a_j s} \check{M}_j H(x, s) ds.$$

Now $\check{M}_j$ acts only on $\gamma(x', s) - i x_0 s$ and leaves $A(s)$ unchanged. Consequently

$$(5.54) \qquad \check{M}_j H(x, s) = O(e^{-\alpha s}), \quad s \to \infty.$$

Thus (5.53) remains valid in the region

$$-(\alpha + 2 \min_j a_j) < \operatorname{Re}\lambda \leq 0, \quad \lambda \neq -\alpha.$$

We continue this process as in (4.46). At the $m$-th step $k_\lambda(x)$ is extended to the region

$$(5.55) \qquad -(\alpha + 2m \min_j a_j) < \operatorname{Re}\lambda \leq 0. \quad \lambda \notin \Lambda.$$

Again with $J = (j_1, \ldots, j_m)$, $1 \leq j_k \leq 2n$ and $|J| = m$ set

$$(5.56) \qquad \lambda_J = \prod_{k=0}^{m} \left( \lambda + \alpha + 2 \sum_{l \leq k} a_{j_l} \right),$$

$$(5.57) \qquad D_J = \lambda_J^{-1} \exp\left( -2 \sum_{l=1}^{m} a_{j_l} s \right) \check{M}_{j_1} \ldots \check{M}_{j_m}.$$

(5.58) THEOREM: *Let $m$ denote an arbitrary positive integer. Then $k_\lambda(x)$ has the following extension to the region (5.55):*

$$k_\lambda(x) = \int_{-\infty + i\varepsilon\,\mathrm{sgn}\,x_0}^{i\varepsilon\,\mathrm{sgn}\,x_0} e^{-\lambda s} H(x,s)\,ds$$

$$(5.59) \qquad\qquad + \sum_{|J| \leq m} D_J H(x, i\varepsilon\,\mathrm{sgn}\,x_0)$$

$$+ \int_{i\varepsilon\,\mathrm{sgn}\,x_0}^{\infty + i\varepsilon\,\mathrm{sgn}\,x_0} e^{-\lambda s} \left( \frac{\partial}{\partial s} + \alpha \right) \sum_{|J|=m} D_J H(x,s)\,ds,$$

*where $0 < \varepsilon < \frac{1}{2} i\pi$ and $H(x,s)$ is defined by (5.33) and (5.44).*

When $\lambda > 0$, we note that (5.25) implies

$$(5.60) \qquad k_\lambda(x_0, x') = k_{-\lambda}(-x_0, x').$$

Substituting this into (5.59) yields the analytic continuation of $k_\lambda(x)$ for $\mathrm{Re}\,\lambda > 0$, $\lambda \notin \Lambda$.

(5.61) REMARK: The case $d = 2n$ and $a_1 = \cdots = a_{2n} = a > 0$ is discussed in Folland and Stein [1], where the group is the Heisenberg group with standard coordinates. In this case (5.41) becomes

$$(5.62) \qquad k_\lambda(x) = \frac{1}{2\pi} \Gamma(n)(4\pi)^{-n} \int_{-\infty}^{\infty} e^{-\lambda s} B(x,s)\,ds,$$

$$(5.63) \qquad B(x,s) = \left[ \frac{1}{4}|x'|^2 \cosh(as) - ia^{-1}x_0 \sinh(as) \right]^{-n}.$$

To evaluate (5.62) when $x' \neq 0$, we set

$$(5.64) \qquad \rho = \left( \frac{1}{16}|x'|^4 + a^{-2}x_0^2 \right)^{\frac{1}{4}}$$

(5.65)      $e^{-i\beta} = \rho^{-2}\left(\frac{1}{4}|x'|^2 - ia^{-1}x_0\right), \quad \beta \in \left(-\frac{1}{2}\pi, \frac{1}{2}\pi\right).$

Then the identity $\cosh(as + i\beta) = \cosh(as)\cos\beta + i\sinh(as)\sin\beta$ gives

(5.66)                    $B(x,s) = [\rho^2 \cosh(as + i\beta)]^{-n}.$

Changing the contour, (5.62) becomes

(5.67)      $k_\lambda(x) = \frac{1}{2\pi}\Gamma(n)(4\pi\rho^2)^{-n}e^{i\lambda a^{-1}\beta}\int_{-\infty}^{\infty} e^{-\lambda s}[\cosh as]^{-n}\,ds.$

This last integral can be evaluated:

(5.68)   $k_\lambda(x) = \frac{1}{4\pi a}(2\pi\rho^2)^{-n}e^{i\lambda a^{-1}\beta}\Gamma\left(\frac{1}{2}[n - a^{-1}\lambda]\right)\Gamma\left(\frac{1}{2}[n + a^{-1}\lambda]\right).$

This extends immediately from the original region $|\operatorname{Re}\lambda| < na$ to the region $C\backslash\Gamma$,

(5.69)                $\Gamma = \{\pm(n + 2m)a, \quad m = 0, 1, 2, 3, \ldots\}.$

If we take $\lambda = i$ and let $a$ tend to zero, $P$ becomes the heat operator, $-\sum_{j=1}^{d}\partial_j^2 + \partial_0$. It is not immediately apparent that $k_i$ in (5.68) tends to the heat kernel. To see that it does note that for $x_0 \neq 0$,

$$e^{i\beta} = -i\operatorname{sgn}x_0 + \frac{1}{4}a|x'|^2|x_0|^{-1} + O(a^2)$$

so

(5.70)                $\beta = -\frac{1}{2}\pi\operatorname{sgn}x_0 + \frac{1}{4}a|x'|^2|x_0|^{-1} + O(a^2).$

Then the identity

$$\Gamma(z)\Gamma(n - z) = (1 - z)(2 - z)\cdots(n - 1 - z)\pi(\sin\pi z)^{-1}$$

with $z = \frac{1}{2}(n - ia)^{-1}$ implies

(5.71)   $a^{n-1}\Gamma\left(\frac{1}{2}[n - a^{-1}i]\right)\Gamma\left(\frac{1}{2}[n + a^{-1}i]\right) = 2^{-n+2}\pi e^{-\frac{1}{2}(\pi/a)} + O(a).$

As $a \to 0$, $a\rho^2 \to |x_0|$ and, with $\lambda = i$, (5.67) converges to

(5.72)           $\frac{1}{2}[1 + \operatorname{sgn}x_0](4\pi|x_0|)^{-n}\exp\left(-\frac{1}{4}|x'|^2|x_0|^{-1}\right).$

Thus we have derived (5.43) in the case $d$ is even and $\lambda = i$.

To derive (5.43) for general $\lambda \in \mathbf{C} \backslash \mathbf{R}$, in a similar way, we need to evaluate
(5.73)

$$\lim_{a \to 0} \frac{1}{2\pi} \Gamma\left(\frac{d}{2}\right) (4\pi)^{-\frac{d}{2}} \int_{-\infty}^{\infty} e^{-\lambda s} \left[\frac{1}{4}|x'|^2 \cosh(as) - i\frac{x_0}{a}\sinh(as)\right]^{-\frac{d}{2}} ds$$

$$= \lim_{a \to 0} \frac{1}{2\pi} (4\pi a\rho^2)^{-\frac{d}{2}} 2^{\frac{d}{2}-1} e^{i\frac{\lambda}{a}\beta} a^{\frac{d}{2}-1} \Gamma\left(\frac{d}{4} - \frac{\lambda}{2a}\right) \Gamma\left(\frac{d}{4} + \frac{\lambda}{2a}\right).$$

If $d = 2n$, a calculation similar to the one that led to (5.71) yields

(5.74)
$$a^{n-1} \Gamma\left(\frac{n}{2} - \frac{\lambda}{2a}\right) \Gamma\left(\frac{n}{2} + \lambda 2a\right)$$
$$= 2\pi \left(\frac{\lambda}{2}\right)^{n-1} [-i \operatorname{sgn}(\operatorname{Im}\lambda)]^{n-1} e^{i\frac{\pi\lambda}{2a}\operatorname{sgn}(\operatorname{Im}\lambda)} + O(a).$$

Now (5.70), (5.73) and (5.74) give (5.43) when $d = 2n$.

If $d = 2n + 1$ we use Stirling's formula for $\Gamma(z)$:

(5.75)
$$a^{\frac{d}{2}-1} \Gamma\left(\frac{d}{4} - \frac{\lambda}{2a}\right) \Gamma\left(\frac{d}{4} + \frac{\lambda}{2a}\right)$$
$$= a^{n-1} \left(\frac{n}{2} - \frac{\lambda}{2a}\right) \left(\frac{n}{2} + \frac{\lambda}{2a}\right)$$
$$\cdot a^{\frac{1}{2}} \exp\left[\frac{1}{4}\log\left(\frac{n}{2} - \frac{\lambda}{2a}\right) + \frac{1}{4}\log\left(\frac{n}{2} + \frac{\lambda}{2a}\right)\right] + O(a).$$

Then

(5.76)
$$\sqrt{a}\exp\left[\frac{1}{4}\log\left(\frac{n}{2} - \frac{\lambda}{2a}\right) + \frac{1}{4}\log\left(\frac{n}{2} + \frac{\lambda}{2a}\right)\right] = \sqrt{a}\, e^{-\frac{\pi}{4}\operatorname{sgn}(\operatorname{Im}\lambda)}$$
$$\cdot \exp\left[\frac{1}{4}\log\left(\frac{\lambda}{2a} - \frac{n}{2}\right) + \frac{1}{4}\log\left(\frac{\lambda}{2a} + \frac{n}{2}\right)\right]$$
$$= \sqrt{a}\, e^{-i\frac{\pi}{4}\operatorname{sgn}(\operatorname{Im}\lambda) + \frac{1}{2}\log\frac{\lambda}{2a}} + O(a)$$
$$= [-i\lambda \operatorname{sgn}(\operatorname{Im}\lambda)]^{\frac{1}{2}} 2^{-\frac{1}{2}} + O(a).$$

Finally (5.70), (5.73), (5.75) and (5.76) imply (5.43) when $d = 2n + 1$.

§6 *Inverting $P_\lambda$ in Skew-Symmetric Form: Symbols*

Recall the differential operator (1.1)

$$P = - \sum_{j,k=1}^{d} g_{jk} Y_j Y_k - iT + c,$$

where $g = (g_{jk})$ is positive definite symmetric at every point $y \in M$. Let $\mathcal{V}$ denote the subbundle of $TM$ generated by $Y_1, \ldots, Y_d$. Let $\langle \, , \, \rangle$ denote an inner product on $\mathcal{V}$ defined by

(6.1) $$\langle Z, Z \rangle = \sum_{j,k=1}^{d} q_j (g^{-1})_{jk} q_k,$$

where

(6.2) $$g^{-1} g = I$$

and

(6.3) $$Z = \sum_{k=1}^{d} q_k Y_k.$$

Then the vector fields

$$X_j = \sum_{k=1}^{d} h_{jk} Y_k, \quad j = 1, \ldots, d,$$

introduced in (1.3) are orthonormal. Here $h = \sqrt{g}$. This follows from the definition:

$$\langle X_m, X_n \rangle = \sum_{j,k=1}^{d} h_{mj} (g^{-1})_{jk} h_{nk}$$
$$= (h g^{-1} h^t)_{mn}$$
$$= (\sqrt{g} \, g^{-1} \sqrt{g})_{mn}$$
$$= \delta_{mn}.$$

Let $X_0$ denote a vector field linearly independent from $\mathcal{V}$. Then

$$P = - \sum_{j=1}^{d} X_j^2 - i\lambda X_0 + \sum_{j=1}^{d} \gamma_j X_j + c_1,$$

see (1.7). Let $C(y) = (c_{jn}(y))$, $y \in M$ denote the skew-symmetric matrix defined by (1.23), i.e.

(6.4) $$[X_k, X_j](y) = c_{jk}(y) X_0(y) \qquad \mod \mathcal{V}_y.$$

Then the model operator in skew-symmetric form is given by

$$(6.5) \qquad P^y = -\sum_{j=1}^{d}\left(\frac{\partial}{\partial x_j} + \frac{1}{2}\sum_{k=1}^{d}c_{jk}(y)x_k\frac{\partial}{\partial x_0}\right)^2 - i\lambda\frac{\partial}{\partial x_0}.$$

(6.6) PROPOSITION: *The model operator (6.5) is independent of the choice of the orthonormal frame $X_1, \ldots, X_d$ of $\mathcal{V}$.*

Proof. Let $Z_1, \ldots, Z^d$ denote another orthonormal frame for $\mathcal{V}$. Then

$$(6.7) \qquad Z_j = \sum_{k=1}^{d} r_{jk}X_k, \quad j = 1, \ldots, d,$$

where $R(y) = (r_{jk}(y))$ is an orthogonal matrix for each $y \in M$. Now

$$\sum_{j=1}^{d}X_j^2 = \sum_{j=1}^{d}\left(\sum_{k=1}^{d}r_{kj}Z_k\right)^2$$

$$(6.8) \qquad = \sum_{k,l=1}^{d}\left(\sum_{j=1}^{d}r_{kj}r_{lj}\right)Z_kZ_l \qquad \text{mod } \mathcal{V}$$

$$= \sum_{k=1}^{d}Z_k^2 \qquad \text{mod } \mathcal{V}.$$

Thus

$$(6.9) \qquad P = -\sum_{j=1}^{d}Z_j^2 - i\lambda X_0 + \sum_{j=1}^{d}\gamma_j'Z_k + c_1.$$

Next

$$[Z_k, Z_j] = \left[\sum_{l=1}^{d}r_{kl}X_l, \sum_{m=1}^{d}r_{jm}X_m\right]$$

$$(6.10) \qquad = \sum_{l,m=1}^{d}r_{kl}r_{jm}[X_l, X_m] \qquad \text{mod } \mathcal{V}$$

$$= \sum_{l,m=1}^{d}r_{kl}r_{jm}(c_{ml}'X_0) \qquad \text{mod } \mathcal{V}$$

$$= (RCR^t)_{jk}X_0 \qquad \text{mod } \mathcal{V}.$$

So the model operator is given by

$$(6.11) \qquad P^y = -\sum_{j=1}^{d}\left(\frac{\partial}{\partial z_j} + \frac{1}{2}\sum_{k=1}^{d}c'_{jk}(y)z_k\frac{\partial}{\partial z_0}\right)^2 - i\lambda(y)\frac{\partial}{\partial z_0}$$

where $C' = RCR^t$. According to (1.28) - (1.30) by changing the variables

$$(6.12) \qquad \begin{cases} x_0 = z_0 \\ x_j = \sum_{k=1}^{d} r_{jk}z_k, \quad j = 1,\ldots,d, \end{cases}$$

(6.11) becomes (6.5). Thus we have derived Proposition 6.6.

In Chapters 4 and 5 we found the symbol and kernel of the inverse of the model operator $P_\lambda$ in normal form. For future applications we need $P_\lambda^{-1}$ when $P_\lambda$ is given in general skew-symmetric form.

In this chapter

$$(6.13) \qquad P_\lambda = -\sum_{j=1}^{d}\left(\frac{\partial}{\partial x_j} + \frac{1}{2}\sum_{k=1}^{d}c_{jk}x_k\frac{\partial}{\partial x_0}\right)^2 - i\lambda\frac{\partial}{\partial x_0},$$

where $C = (c_{jk})$ is a skew-symmetric matrix. We shall construct the symbol of $P_\lambda^{-1}$. (6.13) is left-invariant with respect to the group law

$$(6.14) \qquad x \cdot z = \left(x_0 + z_0 + \frac{1}{2}\sum_{j,k=1}^{d}c_{jk}x_k z_j, \quad x' + z'\right),$$

where $x = (x_0, x')$. Therefore so is $P_\lambda^{-1}$. Thus it suffices to give its symbol at the origin. In Chapter 4 we already found the symbol when $P_\lambda^{-1}$ is in normal form. According to (6.10) and (6.11) an orthogonal change of the variables $x' = (x_1,\ldots,x_d)$ will change $P_\lambda$ from general skew-symmetric form into normal form. Thus all we need to do is to write the formulas of Chapter 4 in a form which is invariant under orthogonal changes of the variable $x'$. For the symbol, at $x = 0$, this is equivalent to saying that it should be invariant under orthogonal changes of $\xi' = (\xi_1,\ldots,\xi_d)$. When $P_\lambda$ is in normal form (4.17) gives

$$\sigma(P_\lambda^{-1})(0,\xi) = \int_0^\infty e^{-s\lambda\xi_0}G(\xi,s)ds$$

where

$$G(\xi,s) = \prod_{j=1}^{d}[\cosh(a_j|\xi_0|s)]^{-\frac{1}{2}}\exp\left[-\sum_{j=1}^{d}\xi_j^2\tanh(a_j|\xi_0|s)(a_j|\xi_0|)^{-1}\right].$$

From (1.26)

$$|A| = \mathrm{diag}(a_1, \ldots, a_n, a_1, \ldots, a_n, 0, \ldots, 0),$$

$a_j > 0$, is the positive square root, $(A^t A)^{\frac{1}{2}}$, of $A^t A$. We write

$$(6.15) \qquad \prod_{j=1}^{d} [\cosh(a_j |\xi_0|s)]^{-\frac{1}{2}} = [\det \cosh(|A||\xi_0|s)]^{-\frac{1}{2}}$$

which is clearly invariant under orthogonal transformations of $A$. In particular we can replace $A$ by $C = (c_{ij})$. Thus the first part of Theorem 4.51 can be rephrased in the following manner

(6.16) THEOREM: *If*

$$(6.17) \qquad -\frac{1}{2} \, \mathrm{trace} \, |C| < \mathrm{Re}\, \lambda < \frac{1}{2} \, \mathrm{trace} \, |C|$$

*or if*

$$(6.18) \qquad \mathrm{Im}\, \lambda \neq 0,$$

*the operator $P_\lambda$ of (6.13) has a two-sided pseudo-differential operator inverse $Q_\lambda$ with symbol*

$$(6.19) \qquad \sigma(Q_\lambda)(x, \xi) = q_\lambda(\sigma_0(x, \xi), \ldots, \sigma_d(x, \xi)),$$

*where $\sigma_j$, $j = 0, 1, \ldots, d$ is the symbol of $i^{-1} X_j$, with $X_j$ given by (3.20), when $b_{jk} = c_{jk}$. The function $q_\lambda$ is given by*

$$(6.20) \qquad q_\lambda(\xi) = \int_o^\infty e^{-\lambda \xi_0 s} G(\xi, s)\, ds$$

*if*

$$(6.21) \qquad -\frac{1}{2} \, \mathrm{trace} \, |C| < \mathrm{Re}\, \lambda < \frac{1}{2} \, \mathrm{trace} \, |C|$$

*where*

$$(6.22) \qquad G(\xi, s) = [\det \cosh(|C||\xi_0|s)]^{-\frac{1}{2}} \exp\left[ -\langle \frac{\tanh(s|C||\xi_0|)}{|C||\xi_0|} \xi', \xi' \rangle \right],$$

*with $\xi = (\xi_0, \xi')$, if $n \neq 0$, or*

$$(6.23) \qquad q_\lambda(\xi) = \left( \lambda \xi_0 + \sum_{j=1}^{d} \xi_j^2 \right)^{-1} \quad if \quad n = 0.$$

*If $0 < 2n < d$ then on the remaining region, $\{\mathrm{Im}\, \lambda \neq 0\}$, $q_\lambda$ is obtained from (6.20) by analytic continuation, by simply changing the ray of integration as in (4.30).*

(6.24) REMARKS:

(i) We note that

(6.25)
$$\frac{\tanh az}{a} = z - a^2 \frac{z^3}{3} + \cdots$$

This defines
$$\frac{\tanh Kz}{K}$$
even for a singular matrix $K$, which justifies (6.22).

(ii) By setting $n = 0$ a formal calculation easily reduces (6.20) to (6.23).

When $2n = d$ the extension of the symbol of $P_\lambda^{-1}$ to the remaining admissible values of $\lambda$, namely $\lambda \in \mathbf{R}$ and $\lambda \neq \pm \sum_{j=1}^{n}(2\alpha_j + 1)a_j$, $\alpha \in \mathbf{Z}_+^n$, can also be given as a function of $|C|$. We shall realize the products occurring in (4.44) and (4.45) as eigenvalues of tensor products of $|A|$. These are invariant under orthogonal transformations of $A$ so we can replace $A$ by $C$. We note that

(6.26)
$$Q^t C Q = A, \qquad Q \in O(2n),$$

implies that

(6.27)
$$Q^t |C| Q = |A|.$$

Recall that

(6.28)
$$\mu = \lambda \xi_0,$$

and

(6.29)
$$\gamma = \frac{1}{2}|\xi_0| \operatorname{trace} |A| = \frac{1}{2}|\xi_0| \operatorname{trace} |C|.$$

We note that the $2n$ quantities
$$\mu_j = (\mu + \gamma)(\mu + \gamma + 2a_j|\xi_0|), \quad j = 1, \ldots, 2n$$

are eigenvalues of the operator

(6.30)
$$E(E + 2|A| \, |\xi_0|),$$

where we set

(6.31)
$$E = (\mu + \gamma)I,$$

$I$, the identity matrix of $\mathbf{R}^{2n}$. Next, if $J = (j_1, j_2)$, the $2n \times 2n$ quantities
$$\mu_{(j_1,j_2)} = (\mu + \gamma)(\mu + \gamma + 2a_{j_1}|\xi_0|)(\mu + \gamma + 2a_{j_1}|\xi_0| + 2a_{j_2}|\xi_0|)$$

are eigenvalues of the matrix

$$(E(E + 2|A|\,|\xi_0|) \otimes I)(E \otimes I + 2|\xi_0|[|A| \otimes I + I \otimes |A|]),$$

etc. Thus we set

$$(6.32) \qquad\qquad I_1 = I, \quad I_{k+1} = I_k \otimes I,$$

$$(6.33) \qquad\qquad A_1 = |A|, \quad A_{k+1} = A_k \otimes I + I_k \otimes A_1,$$

and

$$(6.34) \quad E_A^{(0)} = \mu + \gamma, \quad E_A^{(k+1)} = (E_A^{(k)} \otimes I)((\mu + \gamma)I_{k+1} + 2|\xi_0|A_{k+1}).$$

Then

$$(6.35) \qquad\qquad \{a_{j_1}|\xi_0| + \cdots + a_{j_k}|\xi_0|; \quad j_i = 1, \ldots, 2n\}$$

are the eigenvalues of $A_k|\xi_0|$, and

$$(6.36) \qquad\qquad \{\mu_J : |J| = k\}$$

are the eigenvalues of $E_A^{(k)}$. All these matrices commute.

Next we consider $M_j$, defined by (4.38).

$$M_j = -\left(\frac{1}{2} a_j |\xi_0| \frac{\partial}{\partial \xi_j} - \xi_j\right)^2$$

$$(6.37) \qquad = -\frac{1}{4} a_j^2 \xi_0^2 \exp\left(|\xi_0|^{-1} \sum_{j=1}^{2n} \frac{\xi_j^2}{a_j}\right) \frac{\partial^2}{\partial \xi_j^2} \exp\left(-|\xi_0|^{-1} \sum_{j=1}^{2n} \frac{\xi_j^2}{a_j}\right)$$

$$= \frac{1}{4} \xi_0^2 \exp\left(\frac{\langle |A|^{-1}\xi', \xi'\rangle}{|\xi_0|}\right) \left(-a_j^2 \frac{\partial^2}{\partial \xi_j^2}\right) \exp\left(-\frac{\langle |A|^{-1}\xi', \xi'\rangle}{|\xi_0|}\right).$$

We note that $\langle |C|^{-1}\xi', \xi'\rangle$ is invariant with respect to orthogonal transformations of $\xi'$. Next let

$$(6.38) \qquad\qquad \nabla_{\xi'} u = \begin{pmatrix} \frac{\partial u}{\partial \xi_1} \\ \vdots \\ \frac{\partial u}{\partial \xi_{2n}} \end{pmatrix},$$

$u \in C_c^\infty(\mathbf{R}^{2n})$, denote the gradient mapping and set

$$(6.39) \qquad\qquad \nabla_A u = |A|\nabla_{\xi'} u.$$

Let $^t\nabla_A$ denote its formal transpose. Set

$$(6.40)\qquad \nabla_A^{(1)} = \nabla_A, \quad \nabla_A^{(k+1)} = \nabla_A \otimes \nabla_A^{(k)}, \quad k = 1, 2, \ldots$$

and define

$$(6.41)$$
$$D_A^{(k)} = \frac{1}{2}|\xi_0|\exp\left(\frac{\langle |A|^{-1}\xi', \xi'\rangle}{|\xi_0|}\right)$$
$$^t\nabla_A^{(k)}(E_A^{(k)})^{-1}e^{-2|\xi_0||A_k|}\nabla_A^{(k)}\frac{1}{2}|\xi_0|\exp\left(-\frac{\langle |A|^{-1}\xi', \xi'\rangle}{|\xi_0|}\right).$$

Comparing (4.45) and (6.41) we have

$$(6.42)\qquad\qquad D_A^{(k)} = \sum_{|J|=k} D_J.$$

Thus we can rewrite (4.46):

$$(6.43)\qquad q_\lambda = \sum_{k \le m} D_A^{(k)}G(\xi, 0) + \int_0^\infty e^{-\mu s}\left(\frac{\partial}{\partial s} + \gamma\right)D_A^{(m)}G(\xi, s)ds.$$

Suppose $z' = RX'$, $R \in O(2n)$, denotes another orthonormal set of coordinates. Then

$$C_{x'} = R^t C_{z'} R, \quad \xi'_{z'} = R\xi'_{x'}$$

and

$$|C_{x'}|\nabla_{\xi'_{x'}} = R^t|C_{z'}|\nabla_{\xi'_{x'}}.$$

*Consequently in another orthonormal coordinate system $D_A^{(k)}$ becomes $D_C^{(k)}$, where we obtain $D_C^{(k)}$ by, simply, replacing $A$ by the appropriate $C$ in the definition of $D_A^{(k)}$.*

Using this notation the last part of Theorem 4.51 can be reformulated as follows.

(6.44) THEOREM: *Suppose $2n = d$ and let $G(\xi, s)$ be given by (6.22). Then, for $m = 1, 2, \cdots$, the formula*

$$(6.45)$$
$$q_\lambda = \sum_{k=1}^m D_C^{(k)}G(\xi, 0)$$
$$+ \int_0^\infty e^{-\lambda \xi_0 s}\left(\frac{\partial}{\partial s} + \frac{1}{2}|\xi_0|\operatorname{trace}|C|\right)D_C^{(k)}G(\xi, s)ds$$

*continues the symbol of $P_\lambda^{-1}$ to the region*

$$(6.46)\qquad -\operatorname{Re}\lambda\xi_0 < \frac{1}{2}|\xi_0|\operatorname{trace}|C| + (m+1)\min_j a_j|\xi_0|, \quad \lambda \notin \Lambda,$$

*where $a_j$, $j = 1, \ldots, 2n$ denote the eigenvalues of $|C|$. In fact (6.45) gives*
$\sigma(P_\lambda^{-1})(0,\xi)$, *hence* $\sigma(P_\lambda^{-1})(x,\xi) = q_\lambda(\sigma(x,\xi))$.

In the rest of §6 we consider the behavior of $P_\lambda = P(\lambda)$ near singular $\lambda$, i.e. $\lambda \in \Lambda$, when $d = 2n$. From (4.15) it follows that $Q(\lambda) = P(\lambda)^{-1}$ has a simple pole at each point of $\Lambda$. Let $\lambda_0 \in \Lambda$. We expand $Q(\lambda)$ about $\lambda_0$.

$$(6.47) \qquad Q(\lambda) = \frac{J(\lambda_0)}{\lambda - \lambda_0} + Q(\lambda_0) + O(|\lambda - \lambda_0|).$$

For $\lambda$ sufficiently near $\lambda_0$, $\lambda \neq \lambda_0$

$$(6.48) \qquad Q(\lambda)P(\lambda) = P(\lambda)Q(\lambda) = I,$$

which yields

$$(6.49) \qquad I = \frac{J(\lambda_0)P(\lambda)}{\lambda - \lambda_0} + Q(\lambda_0)P(\lambda) + O(|\lambda - \lambda_0|).$$

According to (4.1)

$$(6.50) \qquad P(\lambda) = P(\lambda_0) - i(\lambda - \lambda_0)\frac{\partial}{\partial x_0}.$$

We let $\lambda \to \lambda_0$ in (6.49):

$$(6.51) \qquad I = \lim_{\lambda \to \lambda_0} \frac{J(\lambda_0)P(\lambda_0)}{\lambda - \lambda_0} - iJ(\lambda_0)\frac{\partial}{\partial x_0} + Q(\lambda_0)P(\lambda_0).$$

Interchanging $Q(\lambda)$ and $P(\lambda)$ we also have

$$(6.52) \qquad I = \lim_{\lambda \to \lambda_0} \frac{P(\lambda_0)J(\lambda_0)}{\lambda - \lambda_0} - i\frac{\partial}{\partial x_0}J(\lambda_0) + P(\lambda_0)Q(\lambda_0).$$

Thus

$$(6.53) \qquad J(\lambda_0)P(\lambda_0) = P(\lambda_0)J(\lambda_0) = 0,$$

$$(6.54) \qquad I = Q(\lambda_0)P(\lambda_0) - iJ(\lambda_0)\frac{\partial}{\partial x_0} = P(\lambda_0)Q(\lambda_0) - i\frac{\partial}{\partial x_0}J(\lambda_0).$$

Since $\frac{\partial}{\partial x_0}$ commutes with everything

$$(6.55) \qquad i\frac{\partial}{\partial x_0}J(\lambda_0) = iJ(\lambda_0)\frac{\partial}{\partial x_0}.$$

Next we multiply (6.54) by $P(\lambda_0)$ on the left (or on the right):

$$(6.56) \qquad P(\lambda_0)Q(\lambda_0)P(\lambda_0) = P(\lambda_0).$$

Thus

$$(6.57) \qquad [P(\lambda_0)Q(\lambda_0)]^2 = P(\lambda_0)Q(\lambda_0).$$

Similarly

$$(6.58) \qquad \left[-i\frac{\partial}{\partial x_0}J(\lambda_0)\right]^2 = -i\frac{\partial}{\partial x_0}J(\lambda_0).$$

So $P(\lambda_0)Q(\lambda_0)$ and $-i\frac{\partial}{\partial x_0}J(\lambda_0)$ are complementary orthogonal projections mapping $L^2 \to L^2$, where

$$(6.59) \qquad Q(\lambda_0) = \frac{1}{2\pi i}\int_\Gamma \frac{Q(\lambda)}{\lambda - \lambda_0}d\lambda.$$

$$(6.60) \qquad J(\lambda_0) = \frac{1}{2\pi i}\int_\Gamma Q(\lambda)d\lambda.$$

Here $\Gamma$ represents a sufficiently small circle about $\lambda_0$. For future use we collect these calculations as follows.

(6.61) THEOREM: *Let $C$ be nonsingular, i.e. $d = 2n$ and let $\lambda_0 \in \Lambda$. Let $Q(\lambda_0)$ and $J(\lambda_0)$ be given by (6.59) and (6.60). Then, the pseudo-differential operator $-i\frac{\partial}{\partial x_0}J(\lambda_0)$ is a projection with domain equal to the complement of the range of $P(\lambda_0)$ in $L^2$ and the complementary projection is $P(\lambda_0)Q(\lambda_0) = Q(\lambda_0)P(\lambda_0)$, with $Q(\lambda_0)$ the partial inverse to $P(\lambda_0)$.*

The first singular values are $\pm \sum_{j=1}^{n} a_j$. We shall calculate $Q$ and $J$ at these poles of $Q(\lambda)$. A simple way to find the symbol of $J$ is to separate out the pole at $\pm \sum_{j=1}^{n} a_j$ in (4.15). Recall

$$\mu_{2\alpha} + \lambda\xi_0 = |\xi_0|\sum_{j=1}^{2n}\frac{a_j}{2}(4\alpha_j + 1) + \lambda\xi_0$$

$$(6.62) \qquad = |\xi_0|\sum_{j=1}^{n}a_j(2\beta_j + 1) + \lambda\xi_0$$

$$= \left[\lambda + (\mathrm{sgn}\,\xi_0)\sum_{j=1}^{n}a_j(2\beta_j + 1)\right]\xi_0,$$

where $\alpha \in Z_+^{2n}$ or $\beta \in Z_+^{n}$. The first zeroes of (6.62) occur at $\lambda = \pm \sum_{j=1}^{n} a_j$ when $\beta_1 = \cdots = \beta_n = 0$. The residues of (4.15) at those two poles are

$$(6.63) \qquad \sigma\left(J\left(\pm\sum_{j=1}^{n}a_j\right)\right) = \begin{cases} \dfrac{2^n}{\xi_0}\exp\left(-\sum_{j=1}^{2n}\dfrac{\xi_j^2}{a_j|\xi_0|}\right), & \xi_0 \lessgtr 0, \\[4mm] 0 & , \quad \xi_0 \gtrless 0. \end{cases}$$

This remarkably simple formula can already be found in Greiner and Stein [2], when $a_1 = \cdots = a_n = 1$. The projection, $-i\frac{\partial}{\partial x_0}J$ then has the symbol $\xi_0\sigma\left(J\left(\pm\sum\limits_{j=1}^{n}a_j\right)\right)$ at $x = 0$.

$Q\left(\pm\sum\limits_{j=1}^{n}a_j\right)$ has a more complicated symbol and we shall use (4.40) for its calculation. At $x = 0$ the formula

(6.64)
$$\sigma(Q(\lambda)) = \frac{1}{\left(\lambda + \operatorname{sgn}\xi_0 \sum\limits_{j=1}^{n}a_j\right)\xi_0}$$
$$\left[1 + \int_0^\infty e^{-\lambda\xi_0 s}\left(\frac{\partial}{\partial s} + \sum\limits_{j=1}^{n}a_j|\xi_0|\right)G(\xi,s)ds\right],$$

$\lambda \neq \pm\sum\limits_{j=1}^{n}a_j$, makes sense in some neighborhood of $\pm\sum\limits_{j=1}^{n}a_j$.

First we calculate $Q\left(\sum\limits_{j=1}^{n}a_j\right)$. When $\xi_0 > 0$, (6.64) is regular at $\lambda = \sum\limits_{j=1}^{n}a_j$. Therefore, to find $\sigma\left(Q\left(\sum\limits_{j=1}^{n}a_j\right)\right)$ when $\xi_0 > 0$ we substitute $\lambda = \sum\limits_{j=1}^{n}a_j$ into (6.64). An integration by parts then brings (6.64) into the form

(6.65)
$$\sigma\left(Q\left(\sum\limits_{j=1}^{n}a_j\right)\right) = \int_0^\infty e^{-\sum\limits_{j=1}^{n}a_j\xi_0 s}G(\xi,s)ds, \quad \xi_0 > 0.$$

When $\xi_0 < 0$ (6.64) becomes

(6.66)
$$\sigma(Q(\lambda)) = \frac{1}{\left(\lambda - \sum\limits_{j=1}^{n}a_j\right)\xi_0}$$
$$\left[1 + \int_0^\infty e^{-\lambda\xi_0 s}\left(\frac{\partial}{\partial s} - \sum\limits_{j=1}^{n}a_j\xi_0\right)G(\xi,s)ds\right].$$

(6.66) has a pole at $\lambda = \sum_{j=1}^{n} a_j$. Thus, when $\xi_0 < 0$, $\sigma\left(Q\left(\sum_{j=1}^{n} a_j\right)\right)$ is the term of order zero in the expansion of (6.66) about $\lambda = \sum_{j=1}^{n} a_j$:

(6.67)
$$\sigma\left(Q\left(\sum_{j=1}^{n} a_j\right)\right) = -\int_0^\infty se^{-\sum_{j=1}^{n} a_j \xi_0 s}\left(\frac{\partial}{\partial s} - \sum_{j=1}^{n} a_j \xi_0\right) G(\xi, s)\,ds$$

$$= -\int_0^\infty s\frac{\partial}{\partial s}\left[\exp\left(-\sum_{j=1}^{n} a_j \xi_0 s\right) G(\xi, s)\right] ds.$$

(6.67) can be integrated by parts into (6.65) when $\xi_0 > 0$.

A similar calculation yields

(6.68)
$$\sigma\left(Q\left(-\sum_{j=1}^{n} a_j\right)\right) = -\int_0^\infty se^{\sum_{j=1}^{n} a_j \xi_0 s}\left(\frac{\partial}{\partial s} + \sum_{j=1}^{n} a_j \xi_0\right) G(\xi, s)\,ds$$

$$= -\int_0^\infty s\frac{\partial}{\partial s}\left[\exp\left(\sum_{j=1}^{n} a_j \xi_0 s\right) G(\xi, s)\right] ds$$

at $x = 0$. We note that the integrals converge by (4.36) and (4.39).

Continuing these remarks we have proved the following:

(6.69) THEOREM: *When $C$ is nonsingular, the symbol, at $x = 0$, of the partial inverse $Q$ of $P(\pm\frac{1}{2}\,\mathrm{trace}\,|C|)$ is given by*

(6.70)
$$\sigma\left(Q\left(\pm\frac{1}{2}\,\mathrm{trace}\,|C|\right)\right)(0, \xi)$$
$$= -\int_0^\infty s\frac{\partial}{\partial s}\left[\exp\left(\mp\frac{1}{2}\xi_0 s\,\mathrm{trace}\,|C|\right) G(\xi, s)\right] ds,$$

*where $G$ is given by (6.22) and $\xi = (\xi_0, \xi')$.*

*The symbol at $x = 0$ of the projection onto the orthogonal complement of the range of $P(\pm\frac{1}{2}\,\mathrm{trace}\,|C|)$ in $L^2$ is*

(6.71)
$$\xi_0 \sigma\left(J\left(\pm\frac{1}{2}\,\mathrm{trace}\,|C|\right)\right)(0, \xi) = \begin{cases} 2^n \exp\left(-\frac{1}{|\xi_0|}\langle |C|^{-1}\xi', \xi'\rangle\right), & \pm\xi_0 < 0 \\[2ex] 0 & , \pm\xi_0 > 0. \end{cases}$$

§7 *Inverting $P_\lambda$ in Skew-Symmetric Form: Kernels*

Theorem 5.38 can be rewritten in the following form:

(7.1) THEOREM: *Suppose $\lambda \in \mathbf{C}$ satisfies either*

$$(7.2) \qquad |\operatorname{Re}\lambda| < \frac{1}{2}\operatorname{trace}|C|,$$

*or*

$$(7.3) \qquad \lambda \notin \mathbf{R}.$$

*Then there is a contour $I = I(\operatorname{sgn} x_0, \lambda) \subset \mathbf{C}$, such that the function $k_\lambda(x)$, which induces $P_\lambda^{-1}$ via convolution is given by*

$$(7.4) \qquad k_\lambda(x) = \frac{1}{2\pi}\Gamma\left(\frac{d}{2}\right)\int_I A(s)e^{-\lambda s}(\gamma(x',s) - ix_0 s)^{-\frac{d}{2}}ds,$$

*where*

$$(7.5) \qquad A(s) = (4\pi)^{-\frac{d}{2}}(\det(|C|s \, \operatorname{cosech}(|C|s)))^{\frac{1}{2}},$$

$$(7.6) \qquad \gamma(x',s) = \frac{1}{4}\langle|C|s\coth(|C|s)x', x'\rangle.$$

$|C|s\coth(|C|s)$ and $|C|s\operatorname{cosech}(|C|s)$ are both well defined even for singular $C$ – see Remark 6.24(i).

When $d = 2n$ we shall write $k_\lambda(x)$ as a function of $C$ for all other admissible values of $\lambda$ on the real axis, i.e. for $\lambda \in \mathbf{R}\backslash\Lambda$, where $\Lambda$ is given by (4.49). Our technique is similar to that of §6 where we did the same for the symbol.

Suppose $F$ is a function of $x_0 \in \mathbf{R}$ and $x' \in \mathbf{R}^{2n}$ which extends holomorphically in a suitable $x_0$-region in $\mathbf{C}$. Let

$$(7.7) \qquad U^{\pm}F(x_0, x') = F\left(x_0 \pm \frac{1}{4}i\sum_{j=1}^{2n}a_j x_j^2,\, x'\right).$$

Then

$$(7.8) \qquad \frac{\partial}{\partial x_j} + \frac{1}{2}ia_j x_j \frac{\partial}{\partial x_0} = U^{-}\frac{\partial}{\partial x_j}U^{+}.$$

Since also $U^{+}\mathcal{L}, U^{-} = \mathcal{L}$, where $\mathcal{L}$ is defined by (5.51), we rewrite $\check{M}_j$ of (5.50) as

$$(7.9) \qquad \check{M}_j = iU^{-}\frac{\partial}{\partial x_j}\mathcal{L}\frac{\partial}{\partial x_j}U^{+}.$$

We can immediately write $U^{\pm}$ in a form invariant under orthogonal transformations of $x'$, namely

(7.10)
$$U_C^{\pm} F(x_0, x') = F\left(x_0 \pm \frac{1}{4} i \langle |C|x', x' \rangle, \ x'\right).$$

Let $Q$ denote the orthogonal matrix which sends $C$ into normal form, i.e.

(7.11)
$$A = Q^t C Q.$$

We set

(7.12)
$$\nabla_C = Q \nabla_A = Q \nabla$$

where $\nabla_A = \nabla$ denotes the usual gradient in $\mathbf{R}^{2n}$ in the coordinate system $x'$ which normalizes $C$. To focus our attention we recall formula (5.57) which we intend to write as a function of $C$. Namely

$$D_J = \lambda_J^{-1} \exp\left(-2 \sum_{l=1}^{m} a_{j_l} s\right) \check{M}_{j_1} \dots \check{M}_{j_m},$$

where

$$\lambda_J = \prod_{k=0}^{m} \left(\lambda + \alpha + 2 \sum_{l \leq k} a_{j_l}\right),$$

and

(7.13)
$$\alpha = \frac{1}{2} \operatorname{trace} |C|.$$

As in §6 we set

(7.14)
$$I_1 = I, \quad I_{k+1} = I_k \otimes I,$$

(7.15)
$$C_1 = |C|, \quad C_{k+1} = C_k \otimes I + I \otimes C_k,$$

and

(7.16)     $\check{E}_C^{(0)} = \lambda + \alpha, \quad \check{E}_C^{(k+1)} = (\check{E}_C^{(k)} \otimes I)([\lambda + \alpha]I_{k+1} + 2C_{k+1}).$

We also need the tensor products of the gradients

(7.17)
$$\nabla_C^{(1)} = \nabla_C, \quad \nabla_C^{(k+1)} = \nabla_C^{(k)} \otimes \nabla_C,$$

and

(7.18)     $(\mathcal{L}\nabla_C)^{(1)} = \mathcal{L}\nabla_C, \ (\mathcal{L}\nabla_C)^{(k+1)} = (\mathcal{L}\nabla_C)^{(k)} \otimes (\mathcal{L}\nabla_C).$

Now we are ready to define $D_C^{(k)}$:

(7.19)     $D_C^{(k)} = i^k U_C^{-} (^t\nabla_C)^{(k)} (\check{E}_C^{(k)})^{-1} e^{-2C_k s} (\mathcal{L}\nabla_C)^{(k)} U_C^{+}.$

Again we note that

(7.20) $$D_A^{(k)} = \sum_{|J|=k} D_J.$$

Now we can write $k_\lambda(x)$ of Theorem 5.58 as a function of $C$.

(7.21) THEOREM: *Let $m$ denote an arbitrary positive integer. Then the extension of $k_\lambda(x)$ to the region*

(7.22) $$-(\alpha + 2m \min_j a_j) < \operatorname{Re} \lambda \le 0, \quad \lambda \notin \Lambda,$$

*has the following form*

$$k_\lambda(x) = \int_{-\infty + i\varepsilon \operatorname{sgn} x_0}^{i\varepsilon \operatorname{sgn} x_0} e^{-\lambda s} H(x, s) ds$$

(7.23) $$+ \sum_{k=0}^{m} D_C^{(k)} H(x, i\varepsilon \operatorname{sgn} x_0)$$

$$+ \int_{i\varepsilon \operatorname{sgn} x_0}^{\infty + i\varepsilon \operatorname{sgn} x_0} e^{-\lambda s} \left( \frac{\partial}{\partial s} + \frac{1}{2} \operatorname{trace} |C| \right) D_C^{(m)} H(x, s) ds,$$

*where $0 < \varepsilon < \frac{1}{2} i\pi$ and $H(x, s)$ is defined by (5.33) and (5.44). For $\lambda > 0$, $k_\lambda(x)$ can be extended via (5.60).*

Again we may obtain the convolution kernel for the partial inverse and for the projection onto the complement of the range of $P_\lambda$ at the first singular points

(7.24) $$\pm \sum_{j=1}^{n} a_j = \pm \frac{1}{2} \operatorname{trace} |C|,$$

provided $C$ is nonsingular. Using the terminology of Chapter 6 we find the kernel $j$ of $J$ by calculating the inverse Fourier transform of (6.63):

(7.25) $$j\left( \pm \sum_{j=1}^{n} a_j \right) = \mp \frac{a_1 \cdots a_n}{(2\pi)^{n+1}} \frac{(n-1)!}{\left( \sum_{j=1}^{2n} \frac{1}{4} a_j x_j^2 \pm i x_0 \right)^n},$$

and the projection onto the complement of the range of $P_\lambda$ is

(7.26) $$\frac{d}{d(ix_0)} j\left( \pm \sum_{j=1}^{n} a_j \right) = \frac{a_1 \cdots a_n}{(2\pi)^{n+1}} \frac{n!}{\left( \sum_{j=1}^{2n} \frac{1}{4} a_j x_j^2 \pm i x_0 \right)^{n+1}}.$$

To calculate the kernel of the partial inverse

(7.27) $$Q(\lambda) = Q^-(\lambda) + Q^+(\lambda)$$

at $\lambda = -\sum_{j=1}^{n} a_j$, we shall use the analytic continuation of

(7.28) $$k_\lambda(x) = k_\lambda^-(x) + k_\lambda^+(x)$$

to $\mathrm{Re}\,\lambda \leq 0$ as described after formula (5.44). $k_\lambda^-(x)$, given by (5.46), is regular for $\mathrm{Re}\,\lambda \leq 0$, so we substitute $\lambda = -\sum_{j=1}^{n} a_j$ to obtain the kernel of $Q^-\left(-\sum_{j=1}^{n} a_j\right)$:

(7.29) $$k^-\left(-\sum_{j=1}^{n} a_j\right) = \int_{-\infty+i\varepsilon\,\mathrm{sgn}\,x_0}^{i\varepsilon\,\mathrm{sgn}\,x_0} \exp\left(\sum_{j=1}^{n} a_j s\right) H(x, s)\,ds.$$

In $k_\lambda^+(x)$, given by (5.48), the term of order zero in the Taylor expansion about $\lambda = -\sum_{j=1}^{n} a_j$ is

(7.30)
$$k^+\left(-\sum_{j=1}^{n} a_j\right) = -\int_{i\varepsilon\,\mathrm{sgn}\,x_0}^{\infty+i\varepsilon\,\mathrm{sgn}\,x_0} s\exp\left(\sum_{j=1}^{n} a_j s\right)\left(\frac{\partial}{\partial s} + \sum_{j=1}^{n} a_j\right) H(x, s)\,ds$$

$$= -\int_{i\varepsilon\,\mathrm{sgn}\,x_0}^{\infty+i\varepsilon\,\mathrm{sgn}\,x_0} s\frac{\partial}{\partial s}\left[\exp\left(\sum_{j=1}^{n} a_j s\right) H(x, s)\right] ds.$$

Since (7.29) can be integrated by parts into the form (7.30) we have derived

(7.31) $$k\left(-\sum_{j=1}^{n} a_j\right) = -\int_{-\infty+i\varepsilon\,\mathrm{sgn}\,x_0}^{\infty+i\varepsilon\,\mathrm{sgn}\,x_0} s\frac{\partial}{\partial s}\left[\exp\left(\sum_{s=1}^{n} a_j s\right) H(x, s)\right] ds.$$

Finally, (5.60) yields

(7.32) $$k\left(\sum_{j=1}^{n} a_j\right)(x_0, x') = k\left(-\sum_{j=1}^{n} a_j\right)(-x_0, x').$$

(7.33) THEOREM: *Let $C$ be nonsingular. Then the convolution kernel for the partial inverse to $P_\lambda\left(\pm\frac{1}{2}\,\mathrm{trace}\,|C|\right)$ is induced by*

$$k\left(\pm\frac{1}{2}\,\mathrm{trace}\,|C|\right)$$

(7.34)

$$= -\int_{-\infty\mp i\varepsilon\,\mathrm{sgn}\,x_0}^{\infty\mp i\varepsilon\,\mathrm{sgn}\,x_0} s\frac{\partial}{\partial s}\left[\exp\left(\frac{1}{2}\,\mathrm{trace}\,|C|\right)H(\mp x_0, x', s)\right]ds,$$

*where $H$ is given by (5.33) and (5.44).*

   *The convolution kernel of the projection onto the complement of the range of $P\left(\pm\frac{1}{2}\,\mathrm{trace}\,|C|\right)$ in $L^2$ is induced by*

$$\frac{d}{d(ix_0)}j\left(\pm\frac{1}{2}\,\mathrm{trace}\,|C|\right)$$

(7.35)

$$= \frac{\sqrt{\det|C|}}{(2\pi)^{n+1}}\frac{n!}{\left(\frac{1}{4}\langle|C|x', x'\rangle \pm ix_0\right)^{n+1}},$$

*where $x = (x_0, x')$.*

## §8 *Solution of the Model Operator in General Form*

   Let

(8.1)
$$X_0 = \frac{\partial}{\partial x_0}$$

(8.2)
$$X_j = \frac{\partial}{\partial x_j} + \frac{1}{2}\sum_{k=1}^{d}b_{jk}x_k\frac{\partial}{\partial x_0}, \quad j = 1, \ldots, d.$$

   The model operator is

(8.3)
$$P_\lambda = -\sum_{j=1}^{d}(X_j)^2 - i\lambda X_0, \quad \lambda \in \mathbf{C}.$$

   $P_\lambda$ is left-invariant with respect to the group law (1.15). We write

(8.4)
$$b_{jk} = c_{jk} + s_{jk},$$

where

(8.5)
$$c_{jk} = \frac{1}{2}(b_{jk} - b_{kj}),$$

(8.6)
$$s_{jk} = \frac{1}{2}(b_{jk} + b_{kj}).$$

Thus

(8.7)
$$C = -C^t,$$

(8.8)
$$S = S^t.$$

According to (1.21) the quadratic change of variables

(8.9)
$$u_0 = x_0 - \frac{1}{4}\sum_{j,k} s_{jk} x_j x_k,$$

(8.10)
$$u_j = x_j, \quad j = 1, \ldots, d$$

reduces the model operator to skew-symmetric form in which the vector fields $X_0, X_1, \ldots, X_d$ have the form

(8.11)
$$X_0 = \frac{\partial}{\partial u_0},$$

(8.12)
$$X_j = \frac{\partial}{\partial u_j} + \frac{1}{2}\sum_{k=1}^{d} c_{jk}u_k\frac{\partial}{\partial u_0},$$

$j = 1,\ldots,d$, where $c_{jk}$ and $s_{jk}$ are defined by (8.5) and (8.6).

(8.13) THEOREM: *Let $k_\lambda(x) = k_\lambda(x_0, x')$ induce the convolution kernel of $P_\lambda^{-1}$ when $P_\lambda$ is in skew-symmetric form as in Theorems 7.1 and 7.21. Then the inverse of the general model operator $P_\lambda$ of (8.3) is a convolution operator, induced by*

(8.14)
$$k_\lambda\left(x_0 - \frac{1}{4}\sum_{j,k=1}^{d} s_{jk}x_j x_k, x'\right)$$

*with respect to the group law*

(8.15)
$$x \cdot y = \left(x_0 + y_0 + \frac{1}{2}\sum_{j,k=1}^{d} b_{jk}x_k y_j, x' + y'\right).$$

In the rest of this chapter we derive the symbol of $P_\lambda^{-1}$, when $P_\lambda$ denotes the model operator in general form, see (8.1), (8.2) and (8.3). To simplify matters we assume that $2n = d$.

Let

(8.16)
$$\sigma(P_\lambda^{-1})(u,\xi) = q_\lambda(\sigma) = q_\lambda(\sigma_0, \sigma')$$

*denote the symbol of $P_\lambda^{-1}$ when $P_\lambda$ is in skew-symmetric form, see (6.20). In skew-symmetric $u$ coordinates*

(8.17)
$$X_0 = \frac{\partial}{\partial u_0},$$

(8.18)
$$X_j = \frac{\partial}{\partial u_j} + \frac{1}{2}\sum_{k=1}^{2n} c_{jk}u_k\frac{\partial}{\partial u_0},$$

$j = 1, 2,\ldots, 2n$, $\sigma_j = \sigma_j(i^{-1}X_j)$ *denotes the symbol of $i^{-1}X_j$, and the inverse, $P_\lambda^{-1}$, is given by*

(8.19)
$$P_\lambda^{-1}f(u) = (2\pi)^{-2n-1}\int_{\mathbf{R}^{2n+1}}\left\{\int_{\mathbf{R}^{2n+1}} e^{i\langle u-v,\eta\rangle}\right.$$
$$\left. q_\lambda(\eta_0, \sigma'(u,\eta))d\eta\right\}f(v)dv, \ f \in C_c^\infty(\mathbf{R}^{2n+1}).$$

To find the symbol of $P_\lambda^{-1}$ in the original $x$-coordinates we set

(8.20)
$$u = \chi(x),$$

and

(8.21) $$\chi f(x) = f(\chi(x)),$$

where $\chi$ denotes the coordinate change (8.9) and (8.10).

Then

(8.22)
$$P_\lambda^{-1} f(\chi(x)) = (2\pi)^{-2n-1} \int_{\mathbf{R}^{2n+1}} \left\{ \int_{\mathbf{R}^{2n+1}} e^{i\langle u-v,\eta\rangle} \right.$$
$$\left. q_\lambda(\eta_0, \sigma'(u,\eta)) d\eta \right\} f(\chi(y)) dy$$
$$= (2\pi)^{-2n-1} \int_{\mathbf{R}^{2n+1}} \left\{ \int_{\mathbf{R}^{2n-1}} e^{i\langle \chi(x)-\chi(y),\eta\rangle} \right.$$
$$\left. q_\lambda(\eta_0, \sigma'(\chi(x),\eta)) d\eta \right\} f(\chi(y)) dy,$$

where, using (8.9) and (8.10)

(8.23) $$\left| \frac{dv}{dy} \right| = 1.$$

To simplify matters we rewrite (8.9) and (8.10) as

(8.24) $$\chi(x) = \left( x_0 - \frac{1}{4} \sum_{j,k=1}^{2n} s_{jk} x_j x_k, \ x' \right),$$

where $x = (x_0, x')$. Then

(8.25)
$$\chi(x) - \chi(y) = \left( x_0 - y_0 - \frac{1}{4}[\langle Sx', x'\rangle - \langle Sy', y'\rangle], \ x' - y' \right)$$
$$= \left( x_0 - y_0 - \frac{1}{4}\langle S(x'-y'), x'+y'\rangle, \ x' - y' \right)$$
$$= M_{x'+y'}(x - y),$$

where

(8.26) $$\mathbf{M}_{z'} = \begin{pmatrix} 1 & -\frac{1}{4}\sum_{k=1}^{2n} s_{k1} z_k & \cdots & -\frac{1}{4}\sum_{k=1}^{2n} s_{k,2n} z_k \\ 0 & 1 & \cdots & 0 \\ & & \cdots & \\ 0 & 0 & \cdots & 1 \end{pmatrix}.$$

We note that $\sigma_j(x,\eta) = \sigma_j(x',\eta)$, i.e. $\sigma_j$ is independent of $x_0$. Therefore

(8.27)
$$P_\lambda^{-1} f(\chi(x)) = (2\pi)^{-2n-1} \int_{\mathbb{R}^{2n+1}} \int_{\mathbb{R}^{2n+1}} e^{i\langle x-y,\xi\rangle}$$
$$q_\lambda(\xi_0, \sigma'(x', {}^t M_{x'+y'}^{-1}\xi)) f(\chi(y)) d\xi\, dy,$$

where we set

(8.28)
$${}^t M_{x'+y'}\eta = \xi.$$

Again we note that

(8.29)
$$\det M_{z'} = 1.$$

To find the symbol of $P_\lambda^{-1}$ we set

(8.30)
$$g(x) = f(\chi(x))$$

and write $g$ as an inverse Fourier transform:

(8.31)
$$g(y) = (2\pi)^{-2n-1} \int_{\mathbb{R}^{2n+1}} e^{i\langle y,\eta\rangle} \hat{g}(\eta) d\eta.$$

Then multiplying the integrand in (8.27) by $e^{-i\langle x,\eta\rangle}$ yields the symbol of $P_\lambda^{-1}$. According to custom we exchange $\xi$ and $\eta$:

(8.32)
$$\sigma(P_\lambda^{-1})(x,\xi) = (2\pi)^{-2n-1} \int_{\mathbb{R}^{2n+1}} \int_{\mathbb{R}^{2n+1}} e^{-i\langle y-x,\eta-\xi\rangle}$$
$$q_\lambda(\eta_0, \sigma'(x', {}^t M_{x'+y'}^{-1}\eta)) d\eta\, dy.$$

We note that $y_0$ does not occur in $q_\lambda$, so we can use the formula

(8.33)
$$\frac{1}{2\pi} \int_{-\infty}^{\infty} \int_{-\infty}^{\infty} e^{i(y_0-x_0)(\eta_0-\xi_0)} d\eta_0 d\xi_0 = \delta(\eta_0 - \xi_0).$$

*Therefore replacing $\eta_0$ by $\xi_0$ in (8.32) we have $\sigma(P_\lambda^{-1})$ for the model operator $P_\lambda$ in general form:*

(8.34)
$$\sigma(P_\lambda^{-1})(x,\xi) = (2\pi)^{-2n} \int_{\mathbb{R}^{2n}} \int_{\mathbb{R}^{2n}} e^{i\langle y'-x',\eta'-\xi'\rangle}$$
$$q_\lambda(\xi_0, \sigma'(x', {}^t M_{x'+y'}^{-1}(\xi_0,\eta'))) d\eta'\, dy'$$
$$= (2\pi)^{-2n} \int_{\mathbb{R}^{2n}} \int_{\mathbb{R}^{2n}} e^{-\langle z',\eta'-\xi'\rangle}$$
$$q_\lambda(\xi_0, \sigma'(x', {}^t M_{2x'+z'}^{-1}(\xi_0,\eta'))) d\eta'\, dz',$$

*where*

(8.35)
$$^t M_{w'}^{-1}(\xi_0, \xi') = \left( \xi_0, \ \xi_1 + \frac{1}{4} \sum_{k=1}^{2n} s_{k,1} w_k \xi_0, \dots, \right.$$
$$\left. \xi_{2n} + \frac{1}{4} \sum_{k=1}^{2n} s_{k,2n} w_k \xi_0 \right).$$

To continue the calculations we assume that $C = (c_{jk})$ of (8.12) is in normal form, i.e.

(8.36)
$$X_0 = \frac{\partial}{\partial x_0},$$

(8.37)
$$X_j = \frac{\partial}{\partial x_j} - \frac{1}{2} a_j x_{n+j} \frac{\partial}{\partial x_0} + \frac{1}{2} \sum_{k=1}^{2n} s_{j,k} x_k \frac{\partial}{\partial x_0},$$

(8.38)
$$X_{n+j} = \frac{\partial}{\partial x_{n+j}} + \frac{1}{2} a_j x_j \frac{\partial}{\partial x_0} + \frac{1}{2} \sum_{k=1}^{2n} s_{n+j,k} x_k \frac{\partial}{\partial x_0},$$

$j = 1, \dots, n$. In this case, according to (4.17), $q_\lambda$ has the following form at $x = 0$:

(8.39)
$$q_\lambda(\xi) = \int_0^\infty e^{-\lambda \xi_0 t} G(\xi, t) dt,$$

*where*

(8.40)
$$G(\xi, t) = \prod_{j=1}^{2n} \cosh(a_j |\xi_0| t)^{-\frac{1}{2}} e^{-\sum_{j=1}^{2n} \delta_j \xi_j^2},$$

*with*

(8.41)
$$\delta_j = \frac{\tanh(a_j |\xi_0| t)}{a_j |\xi_0|}.$$

With this assumption we simplify (8.34) by calculating

(8.42)
$$(2\pi)^{-2n} \int_{\mathbf{R}^{2n}} \int_{\mathbf{R}^{2n}} e^{i(z', \eta' - \xi')} e^{-\sum_{j=1}^{n} \delta_j \sigma_j^2} d\eta' dz',$$

*where*

(8.43)
$$\sigma_j = \sigma_j(\xi_0, {}^t M_{2x'+z'}^{-1}(\xi_0, \eta')),$$

with

(8.44)
$$\sigma_j(\eta) = \eta_j - \frac{1}{2} a_j x_{n+j} \eta_0,$$

(8.45)
$$\sigma_{n+j}(\eta) = \eta_{n+j} + \frac{1}{2} a_j x_j \eta_0,$$

$j = 1, \ldots, n$. First we note that

(8.46)
$$\sigma_j(x', {}^t M_{2x'+z'}^{-1}(\xi_0, \eta')) = \eta_j + \frac{1}{4} \sum_{j=1}^{2n} s_{kj} z_k \xi_0 + \omega_j \xi_0,$$

$j = 1, \ldots, 2n$, where we set

(8.47)
$$\omega_j = -\frac{1}{2} a_j x_{n+j} + \frac{1}{2} \sum_{k=1}^{2n} s_{kj} x_k,$$

(8.48)
$$\omega_{n+j} = \frac{1}{2} a_j x_j + \frac{1}{2} \sum_{k=1}^{2n} s_{k,n+j} x_k,$$

$j = 1, \ldots, n$. To find (8.42) we calculate the $\eta'$ integral first. To do this replace $\eta'$ by $\sigma'$ :

(8.49)
$$(2\pi)^{-2n} \int_{\mathbf{R}^{2n}} e^{i\langle z', \eta' \rangle} e^{-\sum_{j=1}^{2n} \delta_j \sigma_j^2} d\eta'$$

$$= (2\pi)^{-2n} \prod_{j=1}^{2n} e^{iz_j \left( -\frac{1}{4} \sum_{k=1}^{2n} s_{kj} z_k - \omega_j \right) \xi_0} \int_{-\infty}^{\infty} e^{iz_j \sigma} e^{-\delta_j \sigma^2} d\sigma$$

$$= \frac{\pi^n}{(2\pi)^{2n}} e^{-i\left( \frac{1}{4} \langle Sz', z' \rangle + \langle z', \omega' \rangle \right) \xi_0} \frac{e^{-\sum_{j=1}^{2n} \frac{z_j^2}{4\delta_j}}}{(\delta_1 \ldots \delta_{2n})^{\frac{1}{2}}}.$$

Thus we are left with calculating the $z'$-integral:

(8.50)
$$\frac{\pi^n}{(2\pi)^{2n}} \int_{\mathbf{R}^{2n}} e^{-i\langle z', \xi' + \omega' \xi_0 \rangle} e^{-\frac{1}{4} i \langle Sz', z' \rangle \xi_0} \frac{e^{-\sum_{j=1}^{2n} \frac{z_j^2}{4\delta_j}}}{(\delta_1 \ldots \delta_{2n})^{\frac{1}{2}}} dz'$$

$$= \frac{\pi^n}{(2\pi)^{2n}} \int_{\mathbf{R}^{2n}} e^{-i\langle z', \sigma'(X) \rangle} e^{-\frac{1}{4} i \langle Sz', z' \rangle \xi_0 - \sum_{j=1}^{2n} \frac{z_j^2}{4\delta_j}} \frac{dz'}{(\delta_1 \ldots \delta_{2n})^{\frac{1}{2}}}$$

where we set

(8.51) $$\xi_j + \omega_j \xi_0 = \sigma(X_j)(x, \xi),$$

$j = 1, \ldots, 2n$. We change the variable of integration:

(8.52) $$z' = 2\Delta^{\frac{1}{2}} y',$$

where $\Delta^{\frac{1}{2}}$ denotes the positive square root of

(8.53) $$\Delta = \operatorname{diag}(\delta_1, \ldots, \delta_{2n}).$$

Then

$$(8.50) = \frac{1}{\pi^n} \int_{\mathbf{R}^{2n}} e^{-i\langle y', 2\Delta^{\frac{1}{2}} \sigma'\rangle} e^{-i\langle \Delta^{\frac{1}{2}} S\Delta^{\frac{1}{2}} y', y'\rangle \xi_0} e^{-|y'|^2} dy'$$

We choose $\Theta \in O(2n)$, such that

(8.54) $$\Theta^t(\Delta^{\frac{1}{2}} S\Delta^{\frac{1}{2}})\Theta = d = \operatorname{diag}(d_1, \ldots, d_{2n}),$$

and set $y' = \Theta w'$. Then

$$(8.50) = \frac{1}{\pi^n} \int_{\mathbf{R}^{2n}} e^{-i\langle w', 2\Theta^t \Delta^{\frac{1}{2}} \sigma'\rangle} e^{-i\langle dw', w'\rangle \xi_0 - |w'|^2} dw'$$

$$= \prod_{j=1}^{2n} \frac{1}{\sqrt{\pi}} \int_{-\infty}^{\infty} e^{-iw(2\Theta^t \Delta^{\frac{1}{2}} \sigma')_j} e^{-w^2(1 + i\xi_0 d_j)} dw$$

$$= \left[ \prod_{j=1}^{2n} (1 + i\xi_0 d_j)^{-\frac{1}{2}} \right] \exp\left[ -\sum_{j=1}^{2n} \frac{(2\Theta^t \Delta^{\frac{1}{2}} \sigma')_j^2}{4(1 + i\xi_0 d_j)} \right].$$

We note that

(8.55)
$$\prod_{j=1}^{2n} (1 + i\xi_0 d_j)^{-\frac{1}{2}} = [\det(I + i\xi_0 d)]^{-\frac{1}{2}}$$
$$= [\det(I + i\xi_0 \Delta^{\frac{1}{2}} S\Delta^{\frac{1}{2}})]^{-\frac{1}{2}}$$
$$= [\det(I + i\xi_0 \Delta S)]^{-\frac{1}{2}}.$$

Similarly
(8.56)
$$\langle (I + i\xi_0 d)^{-1} \Theta^t \Delta^{\frac{1}{2}} \sigma', \Theta^t \Delta^{\frac{1}{2}} \sigma' \rangle = \langle \Delta^{\frac{1}{2}} (I + i\xi_0 \Delta^{\frac{1}{2}} S\Delta^{\frac{1}{2}})^{-1} \Delta^{\frac{1}{2}} \sigma', \sigma' \rangle$$
$$= \langle (I + i\xi_0 \Delta S)^{-1} \Delta \sigma', \sigma' \rangle.$$

Therefore

$$(8.50) = [\det(I + i\xi_0 \Delta S)]^{-\frac{1}{2}} \exp[-\langle (I + i\xi_0 \Delta S)^{-1} \Delta \sigma', \sigma' \rangle].$$

We collect these formulas.

*When* $X_j$, $0, 1, \ldots, 2n$ *is given by (8.36) - (8.38), the symbol of* $P_\lambda^{-1}$ *is*
$q_\lambda(\sigma) = q_\lambda(\sigma_0, \sigma')$, *where* $\sigma_j = \sigma_j(i^{-1}X_j)$, $j = 0, 1, \ldots, 2n$,

$$(8.57) \qquad q_\lambda(\xi_0, \xi') = \int_0^\infty e^{-\lambda \xi_0 t} G(\xi, t) dt,$$

*with*

$$(8.58) \qquad G(\xi, t) = \det[(I + i\xi_0 \Delta_A(t) S) \cosh(|A||\xi_0|t)]^{-\frac{1}{2}}$$
$$\exp[-\langle (I + i\xi_0 \Delta_A(t) S)^{-1} \Delta_A(t) \xi', \xi' \rangle],$$

*where we set*

$$(8.59) \qquad \Delta_A(t) = \frac{\tanh(|A||\xi_0|t)}{|A||\xi_0|},$$

*and* $A$ *is given by (1.26).*

According to (1.30) $P_\lambda$ is unchanged under orthogonal transformations of the variables $x' \in \mathbf{R}^{2n}$, so this is also true for $P_\lambda^{-1}$. In particular we can replace $A$ by the general skew-symmetric $C$.

We also note that these formulas make sense even when some $a_j \to 0$. Hence we need not assume that $2n = d$. Thus we derived

(8.60) THEOREM: *Let* $X_j$, $j = 0, 1, \ldots, d$ *be defined by (8.1) and (8.2) with* $C = (c_{jk}$ *and* $S = (s_{jk})$ *by (8.5) and (8.6). Suppose*

$$(8.61) \qquad -\frac{1}{2} \operatorname{trace} |C| < \operatorname{Re} \lambda < \frac{1}{2} \operatorname{trace} |C|,$$

*or*

$$(8.62) \qquad \operatorname{Im} \lambda \neq 0.$$

*Then the model operator* $P_\lambda$ *of (8.3) has a two-sided inverse* $P_\lambda^{-1}$ *with symbol*

$$(8.63) \qquad \sigma(Q_\lambda)(x, \xi) = q_\lambda(\sigma),$$

*where*

$$(8.64) \qquad \sigma_j = \sigma(i^{-1} X_j),$$

$j = 0, 1, \ldots, d$. *Furthermore*

$$(8.65) \qquad q_\lambda(\xi) = \int_0^\infty e^{-\lambda \xi_0 t} G(\xi, t) dt,$$

*at least when (8.61) holds, where*

$$(8.66) \qquad G(\xi, t) = \det[I_{C,S} \cosh(|C||\xi_0|t)]^{-\frac{1}{2}}$$
$$\exp[-\langle I_{C,S}^{-1} \Delta_C(t) \xi', \xi' \rangle].$$

*Here*

(8.67)                           $$I_{C,S} = I + i\xi_0 \Delta_C(t)S,$$

*with*

(8.68)                           $$\Delta_C(t) = \frac{\tanh(|C||\xi_0|t)}{|C||\xi_0|}.$$

In the region $\{\text{Im}\,\lambda \neq 0\}$ $q_\lambda$ is obtained from (8.65) by analytic continu-ation, by changing the ray of integration according to (4.30).

(8.69) REMARK: The function $G$ given by (8.66) includes (4.23) and (6.22). Namely we obtain (6.22) by setting $S = 0$ in (8.66) and (4.23) by putting $C$ in normal form. Thus it is appropriate to use the same letter $G$ to designate all three functions.

When $2n = d$ we can analytically extend $q_\lambda$ to $\lambda \in \mathbf{R}$ as long as $\lambda \neq$ $\pm \sum_{j=1}^{n}(2\alpha_j + 1)a_j$, $\alpha \in Z_+^n$, where the $a_j$'s are the eigenvalues of $\frac{1}{2}(B - B^t)$, $B = (b_{jk})$. This extension is carried out by partial integration as it was done in §§4 and 6. To do this we need the analogue of (4.36), namely

(8.70)                 $$\left(\frac{\partial}{\partial t} + \frac{1}{2}\sum_{j=1}^{2n} a_j|\xi_0|\right)G = \sum_{j=1}^{2n} e^{-2a_j|\xi_0|t} M_j G,$$

when $G$ is given by (8.66). To this end we shall calculate the symbol of $M_j G$ when $P_\lambda$ is given by (8.3), in a manner similar to the calculations that led to Theorem 8.60.

Starting from (8.42) with $x = 0$ we calculate

(8.71)
$$(2\pi)^{-2n} \int_{\mathbf{R}^{2n}} \int_{\mathbf{R}^{2n}} e^{i\langle z', \eta' - \xi'\rangle}$$
$$\left[M_l e^{-\sum_{j=1}^{2n} \delta_j \eta_j^2}\right](\xi_0, {}^t M_{z'}^{-1}(\xi_0, \eta'))d\eta'\,dz',$$

where

(8.72)                 $$M_l = -\left(\frac{1}{2}a_l|\xi_0|\frac{\partial}{\partial \eta_l} - \eta_l\right)^2.$$

Then

$$M_l(\xi_0, {}^t M_{z'}^{-1}(\xi_0, \eta')) = -\left(\frac{1}{2}a_l|\xi_0|\frac{\partial}{\partial \eta_l} - \eta_l - \frac{1}{4}\sum_{k=1}^{2n} s_{kl}z_k\xi_0\right)^2.$$

To calculate (8.71) we start with the $\eta'$ integral. We begin with (8.49) by changing variables $\eta' \to \sigma'$:

$$
\begin{aligned}
\text{(8.73)} \quad & (2\pi)^{-2n} \int_{\mathbf{R}^{2n}} e^{i\langle z',\eta'\rangle} \left[ M_l e^{-\sum_{j=1}^{2n} \delta_j \eta_j^2} \right] (\xi_0, {}^t M_{z'}^{-1}(\xi_0, \eta')) d\eta' \\
& = (2\pi)^{-2n} e^{-\frac{1}{4} i \langle Sz', z'\rangle} \int_{\mathbf{R}^{2n}} e^{-\langle z', \sigma'\rangle} M_l(\xi_0, \sigma') e^{-\langle \Delta\sigma', \sigma'\rangle} d\sigma'.
\end{aligned}
$$

where $\Delta$ is given by (8.53),

$$
\sigma' = \eta' + \frac{1}{4}\xi_0 Sz',
$$

and

$$
\text{(8.74)} \qquad M_l(\xi_0, \sigma') = -\left( \frac{1}{2} a_l |\xi_0| \frac{\partial}{\partial \sigma_l} - \sigma_l \right)^2.
$$

To pull $M_l$ outside the Fourier transform we note that

$$
\frac{\partial}{\partial \sigma_l} \to -iz_l, \qquad \sigma_l \to \frac{\partial}{\partial(iz_l)}
$$

outside the integral. Hence

$$
\begin{aligned}
\text{(8.73)} = {} & (2\pi)^{-2n} e^{-\frac{1}{4} i \langle Sz', z'\rangle} M_l'(\xi_0, z') \int_{\mathbf{R}^{2n}} e^{i\langle z', \sigma'\rangle - \langle \Delta\sigma', \sigma'\rangle} d\sigma' \\
= {} & \frac{\pi^n}{(2\pi)^{2n}} e^{-\frac{1}{4}\langle Sz', z'\rangle} M_l'(\xi_0, z') \frac{e^{-\sum_{j=1}^{2n} \frac{z_j^2}{4\delta_j}}}{(\delta_1 \dots \delta_{2n})^{\frac{1}{2}}},
\end{aligned}
$$

where

$$
M_l'(\xi_0, z') = \left( \frac{\partial}{\partial z_j} - \frac{1}{2} a_l |\xi_0| z_l \right)^2.
$$

Thus we are left with calculating the $z'$-integral:

$$\frac{\pi_n}{(2\pi)^{2n}} \int_{\mathbf{R}^{2n}} e^{-i\langle z',\xi'\rangle - \frac{1}{4}i\langle Sz',z'\rangle\xi_0} M_l'(\xi_0,z') \frac{e^{-\sum_{j=1}^{2n} \frac{z_j^2}{4\delta_j}}}{(\delta_1 \ldots \delta_{2n})^{\frac{1}{2}}} dz'$$

$$= \frac{\pi^n}{(2\pi)^{2n}} \int_{\mathbf{R}^{2n}} e^{-i\langle z',\xi'\rangle} \left( \frac{\partial}{\partial z_l} - \frac{1}{2}a_l|\xi_0|z_l + \frac{1}{2}i\xi_0 \sum_{k=1}^{2n} s_{lk}z_k \right)^2$$

(8.75)
$$e^{-\frac{1}{4}i\xi_0\langle Sz',z'\rangle} \frac{e^{-\sum_{j=1}^{2n} \frac{z_j^2}{4\delta_j}}}{(\delta_1 \ldots \delta_{2n})^{\frac{1}{2}}} dz'$$

$$= -\left( \frac{1}{2}a_l|\xi_0|\frac{\partial}{\partial \xi_l} - \xi_l - \frac{1}{2}i\xi_0 \sum_{k=1}^{2n} s_{lk}\frac{\partial}{\partial \xi_l} \right)^2$$

$$\frac{\pi^n}{(2\pi)^{2n}} \int_{\mathbf{R}^{2n}} e^{-i\langle z',\xi'\rangle - \frac{1}{4}i\xi_0\langle Sz',z'\rangle} \frac{\exp\left[ -\sum_{j=1}^{2n} \frac{z_j^2}{4\delta_j} \right]}{(\delta_1 \ldots \delta_{2n})^{\frac{1}{2}}} dz'.$$

Collecting the results obtained so far we have derived

(8.76) LEMMA: *When $G(\xi,t)$ is given by (8.58) the analogue of (4.36) is*

(8.77)
$$\left( \frac{\partial}{\partial t} + \frac{1}{2}\sum_{j=1}^{2n} a_j|\xi_0| \right) G(\xi,t) = \sum_{j=1}^{2n} e^{-2a_j|\xi_0|t} M_j G(\xi,t),$$

*where*

(8.78)
$$M_j = -\left( \frac{1}{2}a_j|\xi_0|\frac{\partial}{\partial \xi_j} - \xi_j - \frac{1}{2}i\xi_0 \sum_{k=1}^{2n} s_{jk}\frac{\partial}{\partial \xi_k} \right)^2.$$

We note that

(8.79)
$$M_j M_k = M_k M_j, \quad j,k = 1,\ldots,2n.$$

To simplify $M_j$ we find a quadratic form

(8.80)
$$\sum_{k,l=1}^{2n} r_{kl}\xi_k\xi_l = \langle R\xi',\xi'\rangle,$$

so that

(8.81)
$$\frac{1}{2}|\xi_0|\left(a_j\frac{\partial}{\partial x_j} - i(\operatorname{sgn}\xi_0)\sum_{m=1}^{2n}s_{jm}\frac{\partial}{\partial\xi_m}\right)e^{-\sum_{k,l=1}^{2n}r_{kl}\xi_k\xi_l}$$
$$= -\xi_j e^{-\sum_{k,l=1}^{2n}r_{kl}\xi_k\xi_l}.$$

In other words

$$\frac{1}{2}|\xi_0|\left(2a_j\sum_{k=1}^{2n}r_{jk}\xi_k - i(\operatorname{sgn}\xi_0)\sum_{m=1}^{2n}s_{jm}\sum_{k=1}^{2n}2r_{mk}\xi_k\right) = \xi_j,$$

or

$$a_j r_{jk} - i(\operatorname{sgn}\xi_0)\sum_{m=1}^{2n}s_{jm}r_{mk} = |\xi_0|^{-1}\delta_{jk}$$
$$\Longleftrightarrow |A|R - i(\operatorname{sgn}\xi_0)SR = |\xi_0|^{-1}I.$$

Therefore

(8.82)
$$R = |\xi_0|^{-1}(|A| - i(\operatorname{sgn}\xi_0)S)^{-1}$$

We note that $|A| \pm iS$ is invertible, since $A$ is invertible and $A$ and $S$ are real. Thus we have the "simplified" version of $M_j$:

(8.83)
$$M_j = -\frac{1}{4}\xi_0^2 e^{\langle R\xi',\xi'\rangle}$$
$$\left(a_j\frac{\partial}{\partial\xi_j} - i(\operatorname{sgn}\xi_0)\sum_{m=1}^{2n}s_{jm}\frac{\partial}{\partial\xi_m}\right)^2 e^{-\langle R\xi',\xi'\rangle}.$$

Next we note that the technique we developed in Chapters 4 and 6 to extend $q_\lambda$ analytically applies in the present situation. Let

(8.84)
$$\nabla_{A,S} : C_c^\infty(\mathbf{R}^{2n}) \to (C_c^\infty(\mathbf{R}^{2n}),\ldots,C_c^\infty(\mathbf{R}^{2n}))$$

denote the mapping whose $j$-th component is

(8.85)
$$a_j\frac{\partial}{\partial\xi_j} - i(\operatorname{sgn}\xi_0)\sum_{m=1}^{2n}s_{jm}\frac{\partial}{\partial\xi_m},$$

$j = 1,\ldots,2n$ and let $^t\nabla_{A,S}$ denote its formal transpose. Then we set
(8.86)
$$D_{A,S}^{(k)} = \frac{1}{2}|\xi_0|\exp\left(\frac{\langle[A - i(\operatorname{sgn}\xi_0)S]^{-1}\xi',\xi'\rangle}{|\xi_0|}\right)$$
$$^t\nabla_{A,S}^{(k)}(E_A^{(k)})^{-1}e^{-2|\xi_0|A_k}\;\nabla_{A,S}^{(k)}\frac{1}{2}|\xi_0|\exp\left(\frac{\langle[A - i(\operatorname{sgn}\xi_0)S]^{-1}\xi',\xi'\rangle}{|\xi_0|}\right),$$

where $E_A^{(k)}$ and $A_k$ are defined by (6.33) and (6.34), respectively, and

$$\nabla_{A,S}^{(1)} = \nabla_{A,S},$$
$$\nabla_{A,S}^{(k+1)} = \nabla_{A,S} \otimes \nabla_{A,S}^{(k)}, \quad k = 1, 2, \ldots$$

$D_{A,S}^{(k)}$ is invariant under orthogonal transformations of $x' \in \mathbf{R}^{2n}$, hence also of $\xi' \in \mathbf{R}^{2n}$ – see the discussion after formula (6.43). Therefore we can replace $A$ by the general real skew-symmetric $C$ and $S$ can be an arbitrary real symmetric matrix. Now the discussion leading to Theorem 6.44 also yields

(8.87) THEOREM: *Let $X_j$, $j = 0, 1, \ldots, 2n$ and $P_\lambda$ be as in Theorem 8.60. Assume $d = 2n$. Then for $m = 1, 2, \ldots$, the formula*

(8.88)
$$q_\lambda(\xi) = \sum_{k=0}^{m} D_{C,S}^{(k)} G(\xi, 0)$$
$$+ \int_0^\infty e^{-\lambda \xi_0 t} \left( \frac{\partial}{\partial t} + \frac{1}{2} |\xi_0| \operatorname{trace} |C| \right) D_{C,S}^{(k)} G(\xi, t) dt$$

*continues $\sigma(P_\lambda^{-1})(0, \xi)$ analytically to the region*

(8.89)       $- \operatorname{Re}(\lambda \xi_0) < \dfrac{1}{2} |\xi_0| \operatorname{trace} |C| + (m + 1) \min\limits_{j} a_j |\xi_0|, \quad \lambda \notin \Lambda$

*where $D_{C,S}^{(k)}$ is given by (8.86) with $A$ replaced by $C$, $G$ is given by (8.66), and the $a_j$, $j = 1, \ldots, 2n$ are the eigenvalues of $|C|$. Finally*

(8.90)                      $\sigma(P_\lambda^{-1})(x, \xi) = q_\lambda(\sigma(x, \xi)),$

*with $\sigma = (\sigma_0, \sigma_1, \ldots, \sigma_{2n})$ and $\sigma_j(x, \xi) = \sigma(i^{-1} X_j)$.*

(8.91) REMARK: For later applications (see §18) it is important to note that $q_\lambda(\xi)$ is a smooth function of $B = (b_{jk})$. This is clear as long as rank $C = 2n$, since $G(\xi, t)$ is an analytic function of $|C|$ and $S$, both of which are smooth functions of $B$.

*When rank $C < 2n$ (see Theorem 8.60) we note that*

(8.92)                      $\dfrac{\tanh(|C| \, |\xi_0| t)}{|C| \, |\xi_0|},$

*and*

(8.93)                      $\cosh(|C| \, |\xi_0| t)$

*are both analytic functions of $|C|^2 = C^t C$, not only of $|C|$. Now $C$, and therefore $C^t C$ are always smooth functions of $B$, and thus so is $q_\lambda(\xi)$.*

# CHAPTER 3

# Pseudodifferential Operators
# on Heisenberg Manifolds

We introduce here a new class of pseudodifferential operators which has a complete asymptotic calculus and which contains the parametrices of the non-elliptic hypoelliptic differential operators $P$ of §1.

## §9 Standard Pseudodifferential Operators

We begin by recalling the basic facts of the standard pseudodifferential operator calculus, both to provide motivation and because we need these facts for later use. Our basic references are Kohn–Nirenberg [1] and Hörmander [1]. A good general introduction is the expository article by L. Nirenberg [1]; for more complete treatment see the recent books by M. Taylor [1] and F. Treves [2].

(9.1) DEFINITION: *Suppose $U$ is an open set in $\mathbf{R}^n$. Given $m \in \mathbf{Z}$ we let $S^m(U)$ denote the following space of functions $q \in C^\infty(U \times \mathbf{R}^n)$:*

*   $q \in S^m(U)$, if for every compact set $K \subset U$ and all multi-indices $\alpha, \beta$, $q$ satisfies the estimates*

$$(9.2) \qquad |D_x^\alpha D_\xi^\beta q(x,\xi)| \leqq C_{K,\alpha,\beta}(1+|\xi|)^{m-|\beta|},$$

*$(x,\xi) \in K \times \mathbf{R}^n$. The index $m$ is the order of the symbol $q$. The corresponding pseudodifferential operator $q(x,D)$ is defined by*

$$(9.3) \qquad q(x,D)\phi(x) = \int e^{i\langle x,\xi\rangle} q(x,\xi)\hat{\phi}(\xi)\mathrm{d}\xi,$$

*where*

$$(9.4) \qquad \mathrm{d}\xi = (2\pi)^{-n}d\xi.$$

Thus

$$(9.5) \qquad q(x,D) : C_c^\infty(U) \to C^\infty(U).$$

*A continuous linear map*

(9.6)
$$T : C_c^\infty(U) \to C^\infty(U)$$

*is said to be smoothing if it extends to a continuous linear map:*

(9.7)
$$T : \mathcal{E}'(U) \to \mathcal{E}(U) = C^\infty(U).$$

*It is well known, and easily verified, that $T$ is smoothing if and only if $T$ has a smooth kernel:*

(9.8)
$$T\phi(x) = \int k(x, y)\phi(y) dy, \quad \phi \in C_c^\infty(U),$$

*where $k \in C^\infty(U \times U)$.*
  *Set*

(9.9)
$$S^{-\infty}(U) = \bigcap_{m \in \mathbf{R}} S^m(U), \quad S^\infty(U) = \bigcup_{m \in \mathbf{R}} S^m(U).$$

(9.10) PROPOSITION: *Suppose that $q \in S^\infty(U)$. Then $Q = q(x, D)$ is smoothing if and only if $q \in S^{-\infty}(U)$.*

Proof. For $q \in S^{-\infty}(U)$ the following integral is absolutely convergent and the order may be changed to give

(9.11)
$$Q\phi(x) = \int e^{i\langle x, \xi \rangle} q(x, \xi) \int e^{-i\langle y, \xi \rangle} \phi(y) dy \, d\xi$$

$$= \int k(x, y)\phi(y) dy,$$

where

(9.12)
$$k(x, y) = \int e^{i\langle x - y, \xi \rangle} q(x, \xi) d\xi$$

is in $C^\infty(U \times U)$.
  Conversely, suppose $Q = q(x, D)$ is smoothing and has kernel $k$ which is in $C^\infty(U \times U)$. For any $\phi \in C_c^\infty(U)$ we have

(9.13)
$$e^{-i\langle x, \xi \rangle} Q(e^{i\langle \cdot, \xi \rangle} \phi)(x) = \int e^{i\langle x, \eta \rangle} q(x, \xi + \eta) \hat{\phi}(\eta) d\eta.$$

Let $q^{(\alpha)} = (\partial/\partial\xi)^\alpha q$. Taylor's formula gives

$$e^{-i\langle x,\xi\rangle} Q(e^{i\langle\cdot,\xi\rangle}\phi)(x) = \sum_{|\alpha|\leq N} \frac{1}{\alpha!} \int e^{i\langle x,\eta\rangle} q^{(\alpha)}(x,\xi)\eta^\alpha \hat{\phi}(\eta)d\eta$$

(9.14)
$$+ r_N(x,\xi,\phi)$$

$$= \sum_{|\alpha|\leq N} \frac{1}{\alpha!} q^{(\alpha)}(x,\xi)D^\alpha\phi(x) + r_N(x,\xi,\phi),$$

$$r_N(x,\xi,\phi)$$

(9.15)
$$= \sum_{|\alpha|=N+1} \frac{N+1}{\alpha!} \int_0^1 \int e^{i\langle x,\eta\rangle} q^{(\alpha)}(x,\xi+t\eta)\eta^\alpha \hat{\phi}(\eta)(1-t)^N d\eta\, dt.$$

If $q \in S^m(U)$, then because $\hat{\phi}$ is rapidly decreasing we may use Peetre's inequality (see Remark 10.76)⁻

(9.16)
$$(1 + |\xi + t\eta|)^s \leq (1 + |\xi|)^s(1 + |t\eta|)^{|s|}$$

to bound the integrands in (9.15) by

(9.17)
$$C_{N,M}(x,\phi)(1 + |\xi|)^{m-N-1}(1 + |\eta|)^{-M}$$

for any $M > 0$. Taking $M > n$ and integrating,

(9.18)
$$|r_N(x,\xi,\phi)| \leq C_N(x,\phi)(1 + |\xi|)^{m-N-1}.$$

In particular, suppose $K \subset U$ is compact, suppose $\phi \in C_c^\infty(U)$ is $\equiv 1$ near $K$, and suppose $x \in K$. Then all but the $\alpha = 0$ term in the sum (9.14) vanish and we have

(9.19)
$$|q(x,\xi) - e^{-\langle x,\xi\rangle} Q(e^{i\langle\cdot,\xi\rangle}\phi)(x)| \leq C_{K,N}(1 + |\xi|)^{-N},$$

for any $N > 0$. Now $k(x,\cdot)\phi(\cdot)$ is in $C_c^\infty(U)$, so the Fourier-transform in the $y$-variable is rapidly decreasing. Thus

(9.20)
$$Q(e^{i\langle\cdot,\xi\rangle}\phi)(x) = \int e^{i\langle y,\xi\rangle} k(x,y)\phi(y)dy$$

is rapidly decreasing, uniformly on $K$. Combining this with (9.19) shows that $q(x,\cdot)$ is rapidly decreasing, uniformly for $x \in K$. To get the result for derivatives, we note that an $x$-derivative $D_j q$ is the symbol of the commutator $[D_j, Q]$, which is also smoothing, while a $\xi$-derivative of $q$ is the symbol of the commutator of $Q$ with multiplication by an $x_j$. Thus the estimates for derivatives follow by induction and we have proved Proposition 9.10.

We want to compose pseudodifferential operators. As defined they map $C_c^\infty(U)$ to $C^\infty(U)$, so the composition need not be defined. *A continuous linear map $T : C_c^\infty(U) \to C^\infty(U)$ is said to be properly supported if*

$$(9.21) \qquad\qquad T : C_c^\infty(U) \to C_c^\infty(U)$$

*and if $T$ extends to a continuous map*

$$(9.22) \qquad\qquad T : C^\infty(U) \to C^\infty(U).$$

*If $T$ is properly supported and $S$ maps $C_c^\infty(U)$ to $C^\infty(U)$, then both $ST$ and $TS$ are well defined as maps from $C_c^\infty(U)$ to $C^\infty(U)$. Modulo smoothing operators, pseudodifferential operators are properly supported.* To prove this we need the following intrinsic characterization of properly supported operators.

(9.23) PROPOSITION: *A continuous linear map $T : C_c^\infty(U) \to C^\infty(U)$ is properly supported if and only if for each compact set $K \subset U$ there are compact set $K', K'' \subset U$, $K \subset K'$, $K \subset K''$, so that*

  (a)  $\operatorname{supp} u \subset K \Longrightarrow \operatorname{supp} Tu \subset K'$,
  (b)  $K'' \cap \operatorname{supp} u = \emptyset \Longrightarrow K \cap \operatorname{supp} Tu = \emptyset$.

Proof. (i) $(9.21) \Longleftrightarrow$ (a).

First (a) clearly implies (9.21). Conversely, we prove that $(9.21) \Longrightarrow$ (a). Suppose it does not. Then there is a sequence $\{u_n; n = 0, 1, 2, \dots\} \subset C_c^\infty(K)$ and a sequence of compact sets $K = K_0 \subset K_1 \subset K_2 \subset \dots$, such that $\bigcup_{j=0}^{\infty} K_j = U$, $\operatorname{supp} Tu_n \not\subset K_n$ but $\operatorname{supp} Tu_n \subset K_{n+1}$. Consequently we can choose a sequence of points $\{x_n; n = 0, 1, 2, \dots\} \subset U$ with $x_n \in K_{n+1} \backslash K_n$ and $Tu_n(x_n) \neq 0$. In particular $x_n \notin \bigcup_{j=1}^{n-1} \operatorname{supp} Tu_j$. If $\varepsilon_n \searrow 0$ sufficiently rapidly, then $u = \sum_{n=0}^{\infty} \varepsilon_n u_n \in C_c^\infty(U)$ while $Tu(x_n) \neq 0$, $n = 0, 1, 2, \dots$. Hence $Tu \notin C_c^\infty(U)$ which contradicts (9.21). Consequently $(9.21) \Longrightarrow$ (a).

(ii) $(9.22) \Longleftrightarrow$ (b).
First, condition (b) $\Longrightarrow$ (9.22):
Choose compact sets $K_1 \subset K_2 \subset K_3 \subset \dots$, $K_n \subset U$, $\bigcup_{j=1}^{\infty} K_j = U$ and let $K_n''$ denote the set corresponding to $K_n$ in (b); we may assume that $K_n'' \subset K_{n+1}''$, $n = 1, 2, 3, \dots$. Choose $\phi_n \in C_c^\infty(U)$ with $\phi_n \equiv 1$ on $K_n''$. Given $u \in C^\infty(U)$ we have $T(\phi_n u) = T(\phi_m u)$ on $K_m$ for all $n \geq m$, since

$\phi_n - \phi_m \equiv 0$ on $K_m''$. Thus $Tu$ may be defined consistently for every $u \in C^\infty(U)$ by setting $Tu = T(\phi_m u)$ on $K_m$. This yields the continuous extension (9.22).

Conversely, suppose (b) fails. Then there are compact sets $K_n'' \subset U$, $n = 0, 1, 2, \ldots$ with $K = K_0'' \subset K_1'' \subset K_2'' \subset \ldots$, $\bigcup_{k=0}^\infty K_k'' = U$ and a sequence of functions $\{u_n, n = 0, 1, 2, \ldots\} \subset C^\infty(U)$ with supp $u_n \cap K_n'' = \emptyset$, such that $Tu_n \not\equiv 0$ on $K$. We may assume that $Tu_n = 1$ somewhere in $K$. Clearly $u_n \to 0$ in $C^\infty(U)$ topology, while $Tu_n \not\to 0$ on $K$. Thus $T$ has no continuous extension to $C^\infty(U)$. This proves Proposition 9.23.

(9.24) REMARK: Thus, roughly speaking, for a properly supported operator $T$, supp $Tu$ doesn't stray too far from supp $u$. This is a natural generalization of supp $Pu \subset$ supp $u$ when $P$ is a differential operator.

(9.25) PROPOSITION: *Suppose $q \in S^m(U)$. Then there is a symbol $q_1 \in S^m(U)$ such that $q_1(x, D)$ is properly supported, while $q_2 = q - q_1$ belongs to $S^{-\infty}(U)$.*

Proof. Define the operator $Q_1$ by

$$(9.26) \qquad Q_1 u(x) = \iint e^{i(x-y,\xi)} q(x,\xi) \chi(x,y) u(y) dy\, d\xi$$

for $u \in C_c^\infty(U)$, where $\chi \in C^\infty(U \times U)$ is chosen so that $\chi \equiv 1$ in a neighborhood of the diagonal, while for each compact set $K \subset U$ the sets

$$(9.27) \qquad K' = \bigcup_{y \in K} \text{supp}\, \chi(\cdot, y), \quad K'' = \bigcup_{x \in K} \text{supp}\, \chi(x, \cdot)$$

are compact. For example, we may choose

$$(9.28) \qquad \chi(x,y) = \sum \phi_j(x) \psi_j(y)$$

where $\{\phi_j\} \subset C_c^\infty(U)$ is a partition of unity subordinate to a locally finite open cover of $U$, $\{\psi_j\} \subset C_c^\infty(U)$, and $\psi_j = 1$ on supp $\phi_j$. It is easily seen that $Q_1$ satisfies (a) and (b) of Proposition 9.23. Plancherel's formula for $u(y)$ and a change of variables gives

$$(9.29) \qquad Q_1 u(x) = \int e^{i(x,\eta) - i(y,\xi-\eta)} q(x,\xi) \chi(x, x+y) \hat{u}(\eta) d\eta\, dy\, d\xi.$$

The integral with respect to $\eta$ and $y$ in (9.29) is absolutely convergent so we may change the order and obtain

$$(9.30) \qquad Q_1 u(x) = \int e^{i(x,\eta)} q(x,\xi) \tilde{\chi}(x, \xi - \eta) \hat{u}(\eta) d\eta\, d\xi$$

where

$$(9.31) \qquad \tilde{\chi}(x,\xi) = \int e^{-i(y,\xi)} \chi(x, x+y) dy.$$

Now $\chi(x, \cdot) \in C_c^\infty(U)$, so $\tilde{\chi}$ is rapidly decreasing in $\xi$. Thus the integral (9.30) is absolutely convergent. This shows that $Q_1 = q_1(x, D)$ with

$$(9.32) \qquad q_1(x, \eta) = \int q(x, \xi + \eta) \tilde{\chi}(x, \xi) d\xi.$$

Taking the Taylor expansion of $q$ around the point $(x, \eta)$ we obtain

$$(9.33) \qquad q_1(x, \eta) = \sum_{|\alpha| \le N} \frac{1}{\alpha!} q^{(\alpha)}(x, \eta) \int \xi^\alpha \tilde{\chi}(x,\xi) d\xi + r_N(x, \eta).$$

The integral indexed by $\alpha$ in (9.33) is just $D_y^\alpha \chi(x, x+y)|_{y=0}$, which vanishes for $\alpha \ne 0$. Thus (9.33) gives

$$(9.34) \qquad q_1(x, \eta) = q(x, \eta) + r_N(x, \eta),$$

$$(9.35)$$

$$r_N(x, \eta) = \sum_{|\alpha| = N+1} \frac{N+1}{\alpha!} \int_0^1 \int q^{(\alpha)}(x, \eta + t\xi) \xi^\alpha \tilde{\chi}(x, \xi) d\xi (1-t)^N dt.$$

Now $q^{(\alpha)} \in S^{m-|\alpha|}(U)$ and $\tilde{\chi}(x, \cdot)$ is of rapid decrease in $\xi$, so it is easily seen that

$$(9.36) \qquad |r_N(x, \eta)| = O((1 + |\eta|)^{m-N-1}),$$

as in the proof of (9.18). The estimates are uniform on compact subsets of $U$, and derivatives of $q - q_1$ are estimated in exactly the same way. This proves Proposition 9.25.

Suppose now that $(q_j)$ is a sequence of symbols with

$$(9.37) \qquad q_j \in S^{m_j}(U), \quad m = m_0 \ge m_1 \ge m_2 \ge \cdots, \ m_j \to -\infty.$$

*A symbol $q \in S^m(U)$ is said to have the asymptotic expansion*

$$(9.38) \qquad q \sim \sum_{j=0}^\infty q_j$$

*if for each integer $N > 0$,*

$$(9.39) \qquad q - \sum_{j < N} q_j \in S^{m_N}(U).$$

(9.40) PROPOSITION: *If $(q_j)$ is a sequence of symbols satisfying (9.37), then there is a symbol $q \in S^m(U)$ having the asymptotic expansion (9.38). Moreover, two such symbols differ by an element of $S^{-\infty}(U)$.*

Proof. We recall Hörmander's proof (see Theorem 2.7 of [1]). Let $K_j$ be an increasing sequence of compact subsets of $U$, so that $\bigcup_j K_j = U$. Choose $\phi \in C_c^\infty(\mathbf{R}^n)$ which vanishes for $|\xi| < \frac{1}{2}$ and is identically 1 for $|\xi| > 1$. We can then select a sequence $t_j \to \infty$ so rapidly that

$$(9.41) \qquad |D_x^\alpha D_\xi^\beta [\phi(\xi/t_j) q_j(x,\xi)]| \le \frac{1}{2^j}(1 + |\xi|)^{m_j-1-|\beta|},$$

$x \in K_i$, $|\alpha| + |\beta| + i \le j$. In fact, since this is only a finite number of conditions for a given $j$, we only need to use the fact that $|\xi|^\gamma |D^\gamma(\phi(\xi/t_j))|$ is uniformly bounded for each $\gamma$. The first statement of Proposition 9.40 now follows by setting

$$(9.42) \qquad q(x,\xi) = \sum_j \phi(\xi/t_j) q_j(x,\xi),$$

and the second is an obvious consequence of (9.39).

For any real $m$, the standard Sobolev space $H^m = H^m(\mathbf{R}^n)$ consists of all tempered distributions $u$ such that the Fourier transform $\hat{u}$ is square integrable and

$$(9.43) \qquad \int (1 + |\xi|^2)^m |\hat{u}(\xi)|^2 d\xi < \infty.$$

The corresponding localized space $H_{\text{loc}}^m(u)$ consists of all distributions $u \in \mathcal{D}'(U)$ with the property that

$$(9.44) \qquad \phi \in C_c^\infty(U) \Longrightarrow \phi u \in H^m(\mathbf{R}^n).$$

$H^m$ is naturally a Hilbert space and $H_{\text{loc}}^m(U)$ is naturally a Frechet space.

(9.45) THEOREM: *If $q \in S^m(U)$ and if $Q = q(x, D)$ is properly supported, then, for every real $s$, $Q$ extends to a continuous map:*

$$(9.46) \qquad Q : H_{\text{loc}}^s(U) \to H_{\text{loc}}^{s-m}(U).$$

Theorem 9.45 is well known (see, for example, Theorem 2 of Nirenberg [1]) and we shall not prove it here.

The following result on the composition of pseudodifferential operators generalizes Proposition 3.10.

(9.47) THEOREM: *Suppose $q_j(x, \xi) \in S^{m_j}(U)$, $j = 1, 2$. If either $Q_1 = q_1(x, D)$ or $Q_2 = q_2(x, D)$ is properly supported, so that the composition $Q_1 Q_2$ is well defined, then $Q_1 Q_2$ is a pseudodifferential operator with symbol $q_1 \circ q_2 \in S^{m_1 + m_2}(U)$. More precisely:*

$$(9.48) \qquad q_1 \circ q_2 \sim \sum_\alpha \frac{1}{\alpha!} \partial_\xi^\alpha q_1 D_x^\alpha q_2.$$

This is a special case of Theorem 2.10 of Hörmander [1]. We note that in (9.48) the terms corresponding to a given multi-index $\alpha$ have order $m_1 + m_2 - |\alpha|$. (9.48) is of interest mainly in case $q_1$ and $q_2$ are homogeneous or are asymptotic sums of homogeneous symbols.

We complete this discussion of pseudodifferential operators in $S^m(U)$ by recalling the construction of the parametrix of an elliptic differential operator. An operator

$$(9.49) \qquad P = P(x, D) = \sum_{|\alpha| \leq m} a_\alpha(x) D^\alpha, \quad a_\alpha \in C^\infty(U)$$

is elliptic if for each $x \in U$ its principal symbol $p_m(x, \xi)$,

$$(9.50) \qquad p_m(x, \xi) = \sum_{|\alpha| = m} a_\alpha(x) \xi^\alpha,$$

is nowhere zero on $\mathbf{R}^n \backslash 0$. Let $\chi \in C^\infty(\mathbf{R}^n)$ vanish near the origin and be $\equiv 1$ near $\infty$. Then

$$(9.51) \qquad q_{-m}(x, \xi) = p_m(x, \xi)^{-1} \chi(\xi)$$

belongs to $S^{-m}(U)$ and $p_m q_{-m} \equiv 1$ for large $\xi$. In view of the expansion (9.48) this means

$$(9.52) \qquad 1 - p \circ q_{-m} = r \in S^{-1}(U),$$

$$(9.53) \qquad 1 - q_{-m} \circ p = s \in S^{-1}(U).$$

To improve this we set

$$(9.54) \qquad \begin{cases} r^{(0)} = 1, \ r^{(k+1)} = r \circ r^{(k)}, \\ r_N = 1 + r^{(1)} + \cdots + r^{(N-1)}. \end{cases}$$

Then

$$(9.55) \qquad (1 - r) \circ r_N = 1 - r^{(N)}, \quad r^{(N)} \in S^{-N}(U).$$

Strictly speaking we should modify $q_{-m}$ so that the corresponding operator is properly supported; then the operators corresponding to $r$ and $s$

are also properly supported, and, in particular, (9.55) makes sense. Such a modification, of course, will not change the symbols' asymptotic behavior as $|\xi| \to \infty$. Now let

$$(9.56) \qquad q_{-m-N} = q_{-m} \circ r^{(-N)}, \; N = 0, 1, 2, \dots .$$

Then

$$(9.57) \qquad \begin{aligned} p \circ (q_{-m} + q_{-m-1} + \cdots + q_{-m-N-1}) = \\ p \circ q_{-m} \circ r_N = (1 - r) \circ r_N = 1 - r^{(N)}. \end{aligned}$$

By Propositions 9.40 and 9.25 we may find a properly supported pseudo-differential operator $Q$ with symbol

$$(9.58) \qquad q \sim \sum_{j=0}^{\infty} q_{-m-j} \in S^{-m}(U).$$

Then $p \circ q - 1 \in S^{-\infty}(U)$, so $PQ - I$ is a smoothing operator. A similar construction starting with $s$ on the left produces a pseudodifferential operator $Q_1$ such that $Q_1 P - I$ is smoothing. Note then that

$$(9.59) \qquad Q_1 - Q = Q_1(I - PQ) - (I - Q_1 P)Q,$$

so $Q_1 - Q$ is smoothing and we may take $Q_1 = Q$ instead. A consequence is the well-known regularity result for elliptic operators:

$$(9.60) \qquad u \in \mathcal{D}'(U), \quad Pu \in C^{\infty}(U) \Longrightarrow u \in C^{\infty}(U).$$

In fact, the statement is local, so we may assume that $u \in \mathcal{E}'(\mathbf{R}^n)$, by using a $C_c^{\infty}$ cut-off function, if necessary. Then

$$(9.61) \qquad u = (I - QP)u + QPu = u_1 + u_2,$$

and $u_1 \in C^{\infty}(U)$ since $I - QP$ is smoothing, while $u_2 \in C^{\infty}(U)$ since $Pu$ is smooth.

*Inverses of the hypoelliptic differential operators $P$ introduced in §1 do not belong to $S^m$.* They are included in the more general class of type $(\frac{1}{2}, \frac{1}{2})$-pseudodifferential operators introduced by Hörmander in [1]. For $m \in \mathbf{Z}$, $S_{\frac{1}{2},\frac{1}{2}}^m(U)$ denotes the space of functions $q \in C^{\infty}(U \times \mathbf{R}^n)$ which satisfy, in place of (9.2), the weaker estimates:

$$(9.62) \qquad |D_x^{\alpha} D_{\xi}^{\beta} q(x, \xi)| \leqq C_{K,\alpha,\beta}(1 + |\xi|)^{m + \frac{1}{2}|\alpha| - \frac{1}{2}|\beta|}, \; x \in K \subset\subset U.$$

*In the formal expansion (9.48), if $q_1$ and $q_2$ are of type $(\frac{1}{2}, \frac{1}{2})$ and of order $m_1$ and $m_2$, then each term of the expansion is of order $m_1 - \frac{1}{2}|\alpha| +$*

$m_2 + \frac{1}{2}|\alpha| = m_1 + m_2$, and the expansion is generally useless. However, we do have the following result:

(9.63) PROPOSITION: *Suppose that either*

$$(9.64) \qquad q_1 \in S^{m_1}(U), \quad q_2 \in S^{m_2}_{\frac{1}{2},\frac{1}{2}}(U),$$

*or*

$$(9.65) \qquad q_1 \in S^{m_1}_{\frac{1}{2},\frac{1}{2}}(U), \quad q_2 \in S^{m_2}(U),$$

*and suppose that one of $Q = q_1(x,D)$ and $Q_2 = q_2(x,D)$ is properly supported. Then $Q_1 Q_2$ has symbol $q_1 \circ q_2 \in S^{m_1+m_2}_{\frac{1}{2},\frac{1}{2}}(U)$. Moreover, the symbol $q_1 \circ q_2$ has the asymptotic expansion (9.48) in the sense that*

$$(9.66) \qquad q_1 \circ q_2 - \sum_{|\alpha|<N} \frac{1}{\alpha!} \partial_\xi^\alpha q_1 D_x^\alpha q_2 \in S^{m_1+m_2-\frac{1}{2}N}_{\frac{1}{2},\frac{1}{2}}(U).$$

The standard proof of Theorem 9.47 will also prove (9.66), and both are special cases of Theorem 2.10' of Hörmander [1]. We shall prove more precise results in §§12-14.

(9.67) REMARKS: (i) Proposition 9.10 on smoothing operators carries over to operators of type $(\frac{1}{2},\frac{1}{2})$ (the argument is similar). Consequently

$$(9.68) \qquad \bigcap_m S^m_{\frac{1}{2},\frac{1}{2}}(U) = S^{-\infty}(U).$$

(ii) Proposition 9.25 on the existence of a properly supported operator corresponding to a given symbol also carries over to $(\frac{1}{2},\frac{1}{2})$-operators with the same proof.

Finally we shall need a standard extension of the celebrated Calderon–Vaillancourt Theorem (see [1]) on the $L^2$-boundedness of $S^0_{\frac{1}{2},\frac{1}{2}}$-operators.

(9.69) THEOREM: *Let $q(x,D)$ be a properly supported operator with symbol in $S^m_{\frac{1}{2},\frac{1}{2}}(U)$. Then for each real $s$, $q(x,D)$ extends to a mapping*

$$(9.70) \qquad q(x,D) : H^{s+m}_{loc}(U) \to H^s_{loc}(U).$$

There are two fundamental difficulties in carrying the parametrix construction for elliptic differential operators over to the operator $P$ of §1 (in the hypoelliptic case). The first is that $P$ is not elliptic: the second order part of its symbol, $p_2$, vanishes on a line in the cotangent bundle for each fixed $x \in U$. Thus we cannot choose $q_{-2}$ as above. As we shall see later on, the construction in §4, carried out for each model operator $P^y$, $y \in U$, does provide us with the symbol of a partial parametrix: an operator $Q_{-2}$, such that $R = PQ_{-2} - I$ and $S = Q_{-2}P - I$ are both operators of type $(\frac{1}{2}, \frac{1}{2})$ and order $-\frac{1}{2}$.

Next, to carry out the construction of a full parametrix for elliptic operators, we used Theorem 9.47. A similar result is unavailable for $(\frac{1}{2}, \frac{1}{2})$-symbols: *there is no effective calculus for the composition of general pseudodifferential operators of type* $(\frac{1}{2}, \frac{1}{2})$, so the symbols $r^{(N)}$, $r_N$ and $q_{-2-N}$, above, remain mysterious – see the discussion leading up to Proposition 9.63.

As we shall see in the next several sections *the operators $Q_{-2}$ and the error operators $R$ and $S$ belong to a more special class for which a full asymptotic calculus does exist.*

## §10 $\mathcal{V}$-Pseudodifferential Operators

We introduce here a class of pseudodifferential operators, a subclass of the operators of type $(\frac{1}{2}, \frac{1}{2})$, which is well adapted to the study of the differential operator $P$ of §1.

(10.1) DEFINITION: *Let $M$ denote a $d+1$ dimensional $C^\infty$ manifold with a "hyperplane bundle" $\mathcal{V}$, i.e. a subbundle $\mathcal{V} \subset TM$, such that for each $y \in M$ the fiber $\mathcal{V}_y$ has codimension one in the tangent space $T_y M$.*

In this section we assume $M = U \subset \mathbf{R}^{d+1}$ and $X_0, X_1, \ldots, X_d$ denote linearly independent vector fields on $U$, such that $X_1, \ldots, X_d \subset \mathcal{V}$.

(10.2) DEFINITION: *$\mathcal{F}_m(U)$, $m \in \mathbf{Z}$ denotes the space of smooth functions $f$, $f \in C^\infty(U \times (\mathbf{R}^{d+1}\backslash(0)))$, which are H-homogeneous of degree $m$ in the last $d+1$ variables:*

$$(10.3) \qquad\qquad f(x, \lambda \cdot \sigma) = \lambda^m f(x, \sigma),$$

$u \in U$, $\sigma \in \mathbf{R}^{d+1}\backslash(0)$, $\lambda > 0$.

Here, $\lambda \cdot \sigma$ denotes the Heisenberg dilation:

$$(10.4) \qquad \lambda \cdot \sigma = \lambda \cdot (\sigma_0, \sigma') = (\lambda^2 \sigma_0, \lambda \sigma'), \ \lambda > 0.$$

(10.5) DEFINITION: *$\mathcal{F}^m(U)$, $m \in \mathbf{Z}$ denotes the subspace of $C^\infty(U \times \mathbf{R}^{d+1})$ consisting of functions $f$ which have an asymptotic expansion:*

$$(10.6) \qquad\qquad f \sim \sum_{j=0}^{\infty} f_{m-j}, \quad f_k \in \mathcal{F}_k(U),$$

*in the sense that for all multi-orders $\alpha, \beta$ and all $N > 0$, we have*

$$(10.7) \qquad \left| D_x^\alpha D_\xi^\beta \left( f - \sum_{j<N} f_{m-j} \right) \right| \leqq C_{\alpha\beta N} \|\xi\|^{m-N-\langle\beta\rangle},$$

*where $C_{\alpha\beta N} = C_{\alpha\beta N}(x)$ is a locally bounded function on $U$,*

$$(10.8) \qquad \|\xi\| = (\xi_0^2 + |\xi'|^4)^{\frac{1}{4}}$$

*and*

$$(10.9) \qquad \langle\beta\rangle = 2\beta_0 + (\beta_1 + \cdots + \beta_d) = \beta_0 + |\beta|.$$

(10.10 PROPOSITION: *Let $f_j \in \mathcal{F}_j(U)$, $j = m, m-1, m-2, \ldots$. Then there exists a function $f \in \mathcal{F}^m(U)$, so that*

$$(10.11) \qquad f \sim \sum_{j=0}^{\infty} f_{m-j}.$$

Proof. The argument is analogous to the proof of Proposition 9.40. In particular we can let $f$ be given by

$$(10.12) \qquad f(x,\xi) = \sum_{j=0}^{\infty} f_{m-j}(x,\xi)\phi(t_j^{-1} \cdot \xi),$$

if $t_j \to \infty$ sufficiently fast.

Next we introduce the symbol class of $\mathcal{V}$-pseudodifferential operators.

(10.13) DEFINITION: $S_{m,\mathcal{V}}(U)$, $m \in \mathbf{Z}$, *denotes the space of functions $q$, $q \in C^\infty(U \times (\mathbf{R}^{d+1}\backslash(0)))$, which have the following form:*

$$(10.14) \qquad \begin{aligned} q(x,\xi) &= f(x, \sigma(x,\xi)) \\ &= f(x, \sigma_0(x,\xi), \ldots, \sigma_d(x,\xi)) \end{aligned}$$

*for some $f \in \mathcal{F}_m(U)$, where*

$$(10.15) \qquad \sigma_j(x,\xi) = \sigma(i^{-1}X_j), \quad j = 0, 1, \ldots, d.$$

(10.16) DEFINITION: $S_\mathcal{V}^m(U)$, $m \in \mathbf{Z}$, *denotes the space of functions $q$, $q \in C^\infty(U \times \mathbf{R}^{d+1})$, which can be put in the following form:*

$$(10.17) \qquad q(x,\xi) = f(x, \sigma(x,\xi))$$

*for some $f \in \mathcal{F}^m(U)$.*

(10.18) DEFINITION: *The symbol class $\bigcup_m S_\mathcal{V}^m(U)$ induces the class of $\mathcal{V}$- (pseudodifferential) operators.*

If $f$ of (10.17) has the expansion

(10.19)
$$f \sim \sum_{j=0}^{\infty} f_{m-j},$$

we set

(10.20)
$$q \sim \sum_{j=0}^{\infty} q_{m-j},$$

where

(10.21)          $q_k(x,\xi) = f_k(x,\sigma(x,\xi)), \quad k = m, m-1, \ldots$

and refer to (10.20) as "*q has asymptotic expansion* $\sum_j q_{m-j}$."

Any monomial $\xi^\alpha$ can be expressed as $\sum a_{\alpha\beta}(x)\sigma(x,\xi)^\beta$ with $a_{\alpha\beta} \in C^\infty(U)$, so every differential operator on $U$ belongs to some $S_{\mathcal{V}}^m(U)$. Next we show that our symbols belong to the class of symbols of type $(\frac{1}{2},\frac{1}{2})$.

(10.22) PROPOSITION: *For* $m \in \mathbf{Z}$:

(10.23)                $S_{\mathcal{F}}^m(U) \subset S_{\frac{1}{2},\frac{1}{2}}^m(U) \quad if \quad m \geq 0,$

*and*

(10.24)                $S_{\mathcal{F}}^m(U) \subset S_{\frac{1}{2},\frac{1}{2}}^{\frac{1}{2}m}(U) \quad if \quad m < 0.$

Proof. Let $q \sim q_m$ where $q_m \in S_{m,\mathcal{F}}(U)$. Set

(10.25)          $\rho(x,\xi) = \left[ 1 + |\sigma_0(x,\xi)| + \sum_{j=1}^{d} \sigma_j(x,\xi)^2 \right]^{\frac{1}{2}}.$

It follows from (10.20) and the homogeneity of the function $f_m \in \mathcal{F}_m(U)$ (which defines $q_m$) that

(10.26)                $|q(x,\xi)| \leq C_0(x)\rho(x,\xi)^m,$

where $C_0$ is locally bounded. Now

(10.27)          $C(x)^{-1}(1 + |\xi|)^{\frac{1}{2}} \leq \rho(x,\xi) \leq C(x)(1 + |\xi|)$

for some locally bounded function $C$ having locally bounded reciprocal. Thus we have the necessary estimate for the symbol $q$ itself. To get estimates for derivatives, note that differentiation of $q_m$ with respect to $\xi$ involves differentiation of the corresponding $f_m$ with respect to $\sigma$, which lowers the degree of homogeneity of $f_m$ to $m - 1$ (for $\partial/\partial\sigma_j$, $j > 0$) or to $m - 2$ (for $\partial/\partial\sigma_0$). Thus one obtains estimates of the desired type for the pure $\xi$-derivatives of any order. Differentiating $q(\sim q_m)$ with respect to $x$ leads to an $x$-derivative of $f_m$ and also to a $\xi$-derivative of $f_m$; the latter lowers homogeneity but introduces a factor which is $O(1 + |\xi|)$. The total effect is dominated, therefore, by $\rho(x,\xi)^{-1}(1 + |\xi|)$ which is dominated by $(1 + |\xi|)^{\frac{1}{2}}$. Continuing,

$$(10.28) \quad \begin{aligned} D_x^\alpha D_\xi^\beta q(x,\xi)| &\leqq C_{\alpha\beta}(x)\rho(x,\xi)^{m-|\beta|}(1 + |\xi|)^{\frac{1}{2}|\alpha|} \\ &\leqq C'_{\alpha\beta}(x)\rho(x,\xi)^m (1 + |\xi|)^{(\frac{1}{2})|\alpha|-\frac{1}{2}|\beta|} \end{aligned}$$

and using (10.25) we obtain Proposition 10.22 when $q \sim q_m$. To finish the proof we need the following characterization of $q \in S_\mathcal{V}^m(U)$:

(10.29) PROPOSITION: *Let* $q \in C^\infty(U \times \mathbf{R}^{d+1})$ *and* $q_{m-j} \in S_{m-j,\mathcal{V}}(U)$, $j = 0, 1, 2, \ldots$. *Then*

$$(10.30) \quad q \sim \sum_{j=0}^\infty q_{m-j},$$

*if and only if for each pair of multi-indices* $\alpha, \beta$ *and every* $M > 0$ *there exists a positive integer* $N$ *and a locally bounded function* $C_{\alpha\beta M}(x)$, *such that*

$$(10.31) \quad \left| D_x^\alpha D_\xi^\beta \left( q(x,\xi) - \sum_{j<N} q_{m-j}(x,\xi) \right) \right| \leqq C_{\alpha\beta M}(x)|\xi|^{-M}$$

*when* $|\xi| > 1$.

Proof. Let $f \in C^\infty(U \times \mathbf{R}^{d+1})$ and $f_{m-j} \in \mathcal{F}_m(U)$, such that

$$(10.32) \quad q(x,\xi) = f(x, \sigma(x,\xi)),$$

and

$$(10.33) \quad q_{m-j}(x,\xi) = f_{m-j}(x, \sigma(x,\xi)), \quad j = 0, 1, 2, \ldots.$$

Suppose (10.30) holds. Then (10.7) and the proof of Proposition 10.22 give

$$(10.34) \quad \begin{aligned} &\left| D_x^\alpha D_\xi^\beta \left( f(x, \sigma(x,\xi)) - \sum_{j<N} f_j(x, \sigma(x,\xi)) \right) \right| \\ &\leq \sum_{\langle\gamma\rangle\leq\langle\alpha\rangle} C_\gamma(x)\|\sigma\|^{m-N-\langle\beta\rangle-\langle\gamma\rangle}|\xi|^{|\gamma|} \\ &\leq \sum_{\langle\gamma\rangle\leq\langle\alpha\rangle} C'_\gamma(x)(1+|\xi|)^m(1+|\xi|)^{-\frac{1}{2}(N+\langle\beta+\gamma\rangle)+|\gamma|} \end{aligned}$$

assuming $|\xi| > 1$. Choosing $N$ sufficiently large we obtain (10.31).

Conversely, suppose (10.31) is satisfied. We shall show that for all multi-indices $\alpha, \beta$ and for every positive integer $K$,

$$(10.35) \quad \left| D_x^\alpha D_\sigma^\beta \left( f(x,\sigma) - \sum_{j<K} f_{m-j}(x,\sigma) \right) \right| \leq C_{\alpha\beta K}(x)(1+\|\sigma\|)^{m-\langle\beta\rangle-K}.$$

We note that

$$(10.36) \qquad\qquad\qquad \sigma = \sigma(x,\xi)$$

is non-singular and has an inverse

$$(10.37) \qquad\qquad\qquad \xi = \xi(x,\sigma)$$

which is linear in $\sigma$. Therefore (10.31) implies the following estimate:

$$(10.38) \quad \begin{aligned} &\left| D_x^\alpha D_\sigma^\beta \left( q(x,\xi(x,\sigma)) - \sum_{j<N} q_{m-j}(x,\xi(x,\sigma)) \right) \right| \\ &\qquad\qquad \leq C_{\alpha\beta M}(x)|\xi|^{-M}|\sigma|^{|\alpha|}, \end{aligned}$$

if $|\xi| > 1$ and $|\sigma| > 1$. Again, $C_{\alpha\beta M}(x)$ is locally bounded on $U$. Consequently

$$(10.39) \quad \begin{aligned} &\left| D_x^\alpha D_\sigma^\beta \left( f(x,\sigma) - \sum_{j<K} f_{m-j}(x,\sigma) \right) \right| \\ &\leq \left| D_x^\alpha D_\sigma^\beta \left( q(x,\xi(x,\sigma)) - \sum_{j<N} q_{m-j}(x,\xi(x,\sigma)) \right) \right| \\ &\quad + \left| D_x^\alpha D_\sigma^\beta \sum_{K\leq j<N} f_{m-j}(x,\sigma) \right| \\ &\leq C_{\alpha\beta M}(x)\left(|\xi|^{-M}|\sigma|^{|\alpha|} + (1+\|\sigma\|)^{m-K-\langle\beta\rangle}\right), \end{aligned}$$

where we used the $H$-homogeneity of the functions $f_k(x,\sigma)$. Choosing $M$, and therefore $N$, sufficiently large and using

$$(10.40) \qquad 1+|\xi| \geq C(1+\|\xi\|) \quad \text{and} \quad |\xi| \sim |\sigma|,$$

we see that $(10.39) \implies (10.35)$. This proves Proposition 10.29.

*End of proof of Proposition 10.22.* Let $q \in S_V^m(U)$. Then
$$(10.41)$$
$$\begin{aligned} |D_x^\alpha D_\xi^\beta q(x,\xi)| &\leq \left| D_x^\alpha D_\xi^\beta \left( q(x,\xi) - \sum_{j<N} q_{m-j}(x,\xi) \right) \right| + \sum_{j<N} |D_x^\alpha D_\xi^\beta q_{m-j}(x,\xi)| \\ &\leq C_{\alpha\beta M}(x)\left((1+|\xi|)^{-M} + \sum_{j<N}(1+|\xi|)^{k_j-\frac{1}{2}|\beta|+\frac{1}{2}|\alpha|}\right), \end{aligned}$$

where $k_j = m - j$ if $m \geq j$ and $k_j = \frac{1}{2}(m - j)$ if $m - j$. Choosing $M$ sufficiently large $(10.41) \Longrightarrow (10.23)$ if $m \geq 0$ and $\Longrightarrow (10.24)$ if $m < 0$, whence we have obtained Proposition 10.22.

When a fixed frame is chosen for $TM$, the terms $q_{m-j}$, $j = 0, 1, 2, \ldots$ in $(10,.20)$ are $H$-homogeneous in $\sigma$ with successively decreasing rates of growth as $|\sigma| \to \infty$, so they are uniquely determined by the symbol $q$.

(10.42) DEFINITION: *The "principal symbol" of a $\mathcal{V}$-operator of degree $m$, i.e. of a pseudodifferential operator induced by $q \in S_{\mathcal{V}}^m(U)$, is the term $q_m$ in the asymptotic expansion of its symbol $q$:*

$$(10.43) \qquad q \sim \sum_{j=0}^{\infty} q_{m-j}.$$

Let $q \in S_{\mathcal{V}}^{\infty}(U) = \bigcup_m S_{\mathcal{V}}^m(U)$. The correspondence between $\xi$ and $\sigma$ given by $(10.36)$ and $(10.37)$ shows that $D_x^\alpha D_\xi^\beta q(x, \xi)$ is rapidly decreasing in $\xi$, for all $\alpha$ and $\beta$, if and only if $D_x^\alpha D_\sigma^\beta f(x, \sigma)$ is rapidly decreasing in $\sigma$, for all $\alpha$ and $\beta$, where

$$(10.44) \qquad q(x, \xi) = f(x, \sigma(x, \xi)).$$

Consequently, in that case $f_{m-j} \equiv 0$, $j = 0, 1, 2, \ldots$ in $(10.6)$, and therefore $q_{m-j} \equiv 0$, $j = 0, 1, 2, \ldots$ in $(10.20)$. Thus $D_x^\alpha D_\xi^\beta q(x, \xi)$ is rapidly decreasing in $\xi$ for all $\alpha$ and $\beta$, if and only if $q \in S_{\mathcal{V}}^{-\infty}(U) = \bigcap_m S_{\mathcal{V}}^m(U)$. Following the proof of Theorem 2.8 of Hörmander [1] we have

(10.45) PROPOSITION: *Let $q \in S_{\mathcal{V}}^{\infty}(U)$. $q$ is smoothing, i.e. $q(x, D)$ has $C^\infty$ distribution kernel if and only if $q \in S_{\mathcal{V}}^{-\infty}(U)$.*

(10.46) PROPOSITION: *The class of $\mathcal{V}$-operators, $Op\, S_{\mathcal{V}}^m(U)$, does not depend on the choice of a frame, $X_0, X_1, \ldots, X_d$, for $TU$, where $X_1, \ldots, X_d$ is a frame for $\mathcal{V}$.*

Proof. Suppose $q \in S_{\mathcal{V}}^m(U)$ such that $q \sim q_m \in S_{m,\mathcal{V}}(U)$, where

$$(10.47) \qquad q_m(x, \xi) = f_m(x, \sigma(x, \xi)), \quad f_m \in \mathcal{F}_m(U).$$

Suppose that $\{\tilde{X}_j, j = 0, 1, \ldots, d\}$ is another set of linearly independent vector fields with $\tilde{X}_1, \ldots, \tilde{X}_d \in \mathcal{V}$. First let

$$(10.48) \qquad X_0 = a\tilde{X}_0, \quad a \in C^\infty(U).$$

Then

(10.49) $$X_j = \sum_{k=1}^{d} A_{jk}\tilde{X}_k, \quad A_{jk} \in C^\infty(U),$$

and

(10.50) $$\begin{aligned} q(x,\xi) &= f(x,\sigma(x,\xi)) \\ &= f(x, a(x)\tilde{\sigma}_0(x,\xi), A(x)\tilde{\sigma}'(x,\xi)) \\ &= g(x, \tilde{\sigma}(x,\xi)), \end{aligned}$$

where $g$, so defined, belongs to $\mathcal{F}^m(U)$, with

(10.51) $$g \sim g_m,$$

and

(10.52) $$g_m(x,\tilde{\sigma}) = f_m(x, a\tilde{\sigma}_0, A\tilde{\sigma}').$$

This shows that $Op\,S_{m,\nu}(U)$, hence $Op\,S_\nu^m(U)$, is independent of the choice of a frame for $TU$ as long as the direction of $X_0$ remains unchanged. We are left with the case

(10.53) $$X_0 = \tilde{X}_0 + l(x,\tilde{X}') \quad \text{and} \quad X_j = \tilde{X}_j, \; j = 1,\ldots,d,$$

where $l(x,\sigma')$ is linear in $\tilde{\sigma}'$, $\tilde{\sigma}' \in \mathbf{R}^d$. Again we assume $f \sim f_m$, $f_m \in \mathcal{F}_m(U)$. Using Taylor's expansion with remainder we have

(10.54) $$\begin{aligned} f_m(x,\sigma) &= f_m(x, \tilde{\sigma}_0 + l(x,\tilde{\sigma}'), \sigma') \\ &= \sum_{k<N} \frac{1}{k!}((\partial_{\sigma_0}^k f_m)(x,\tilde{\sigma}))l(x,\tilde{\sigma}')^k + R_N, \end{aligned}$$

where

(10.55) $$R_N(x,\tilde{\sigma}) = \frac{l(x,\tilde{\sigma}')^N}{(N-1)!} \int_0^1 (\partial_{\sigma_0}^N f_m)(x, \sigma_0 + tl(x,\tilde{\sigma}'), \tilde{\sigma}')(1-t)^{N-1}dt.$$

Now, for $\|\tilde{\sigma}\| \geq 1$,

(10.56) $$\begin{aligned} &(\partial_{\sigma_0}^N f_m)(x, \tilde{\sigma}_0 + tl(x,\tilde{\sigma}'), \tilde{\sigma}') \\ &= \|\tilde{\sigma}\|^{m-2N}(\partial_{\sigma_0}^N f_m)\left(x, \frac{\tilde{\sigma}_0 + tl(x,\tilde{\sigma}')}{\|\tilde{\sigma}\|^2}, \frac{\tilde{\sigma}'}{\|\tilde{\sigma}\|}\right). \end{aligned}$$

Consequently

(10.57) $$|R_N(x,\tilde{\sigma})| \leq C_N(x)(1 + \|\tilde{\sigma}\|)^{m-N},$$

with some locally bounded function $C_N(x)$. Similarly,

(10.58) $$|D_x^\alpha D_{\tilde{\sigma}}^\beta R_N(x,\tilde{\sigma})| \leq C_{\alpha\beta N}(x)(1 + \|\tilde{\sigma}\|)^{m-N-\langle\beta\rangle}.$$

Thus, according to Definition 10.5, we have derived

$$\tilde{f}(x,\tilde{\sigma}) = f(x,\tilde{\sigma}_0 + l(x,\tilde{\sigma}'),\tilde{\sigma}')$$

(10.59)
$$\sim \sum_{k=0}^{\infty} \frac{1}{k!} l(x,\tilde{\sigma}')^k (\partial_{\sigma_0}^k f_m)(x,\tilde{\sigma}) \in \mathcal{F}^m(U).$$

The term with index $k$ in this expansion is $H$-homogeneous in $\tilde{\sigma}$ of degree $(m-2k)+k = m-k$, since differentiation with respect to $\sigma_0$ reduces the degree of $H$-homogeneity by two. It follows readily that $q \sim q_m$ relative to $\sigma(x,\xi)$ also has an asymptotic expansion in $S_\mathcal{V}^m(U)$ relative to $\tilde{\sigma}(x,\xi)$.

Still assuming (10.53), let

(10.60)
$$f(x,\sigma) \sim \sum_j f_{m-j}(x,\sigma).$$

Set

(10.61)
$$\tilde{f}(x,\tilde{\sigma}) = f(x,\tilde{\sigma}_0 + l(x,\tilde{\sigma}'),\tilde{\sigma}')$$

and

(10.62)
$$\tilde{f}_{m-j}(x,\tilde{\sigma}) = \sum_{l+k=j} \frac{1}{k!} \, l(x,\tilde{\sigma}')^k (\partial_{\sigma_0}^k f_{m-l})(x,\tilde{\sigma}).$$

Clearly

(10.63)
$$\tilde{f}(x,\tilde{\sigma}) \sim \sum_{j=0}^{\infty} \tilde{f}_{m-j}(x,\tilde{\sigma}).$$

Therefore, $q \in S_\mathcal{V}^m(U)$ with an asymptotic expansion (10.20) relative to the $\sigma$-frame for $\mathcal{V}$ has a similar expansion relative to the $\tilde{\sigma}$-frame for $\mathcal{V}$, namely

(10.64)
$$\tilde{q} \sim \sum_{j=0}^{\infty} \tilde{q}_{m-j},$$

where

(10.65)
$$\tilde{q}_k(x,\xi) = \tilde{f}_k(x,\tilde{\sigma}(x,\xi)),$$

$k = m, m-1, m-2, \ldots$ This proves Proposition 10.46.

(10.66) REMARK: To obtain a simple example of this result we introduce a new frame for $TU$ by $\tilde{X}_0 = X_0 + aX_1$, $a \neq 0$, $\tilde{X}_j = X_j$, $j = 1, \ldots, d$. Then the $\mathcal{V}$-operator $X_0$, which has one term in its asymptotic expansion with respect to the original frame, will have two nonzero terms in its asymptotic expansion with respect to the new frame.

A $\mathcal{V}$-frame of $TU$ is a basis $\{X_j, \quad j = 0, 1, \ldots, d\}$ of $TU$ such that $\{X_1, \ldots, X_d\} \subset \mathcal{V}$. We have proved:

"Given a hyperplane bundle $\mathcal{V} \subset TU$, the class of $\mathcal{V}$-operators depends only on $\mathcal{V}$ and not on the choice of $\mathcal{V}$-frame for $TU$."

The next result shows that the class of $\mathcal{V}$-operators is invariant under coordinate transformations.

(10.67) INVARIANCE THEOREM: *Suppose* $U_j, j = 1, 2$, *is an open subset of* $\mathbf{R}^{d+1}$, *and suppose* $\phi : U_1 \to U_2$ *is a diffeomorphism onto. Suppose* $\mathcal{V}_2$ *is a hyperplane bundle on* $U_2$ *and let* $\mathcal{V}_1$ *denote its pullback to* $U_1$. *Given a* $\mathcal{V}_2$-*operator* $Q_2$ *acting on* $\mathcal{E}'(U_2)$, *let* $Q_1$ *be its pullback to* $U_1$:

$$(10.68) \qquad Q_1(u \circ \phi) = (Q_2 u) \circ \phi, \quad u \in \mathcal{E}'(U_2).$$

*Then* $Q_1$ *is a* $\mathcal{V}_1$-*operator.*

The Invariance Theorem implies that $\mathcal{V}$-*operators may be defined on any smooth manifold* $M$ *carrying a hyperplane bundle* $\mathcal{V}$. We postpone its proof to §16.

In Proposition 3.15 we had already discussed an important special case of the Invariance Theorem, namely, when $\phi$ is an affine map, $\phi(x) = a + Ax$, $a \in U_1$ and $A \in \mathrm{GL}(d+1)$. If $q_j(x, \xi)$ denotes the symbol of $Q_j, \quad j = 1, 2$, then

$$(10.69) \qquad q_1(x, \xi) = q_2(\phi(x), (d\phi^{-1})^t \xi).$$

In particular, if $X_j^{(1)}, \quad j = 0, 1, \ldots, d$ denotes the pullback of the $\mathcal{V}_2$-frame $X_j^{(2)}, \quad j = 0, 1, \ldots, d$ from $TU_2$ to $TU_1$, then

$$\begin{aligned} \sigma_j^{(1)}(x, \xi) &= \sigma(i^{-1} X_j^{(1)})(x, \xi) \\ &= \sigma(i^{-1} X_j^{(2)})(\phi(x), (d\phi^{-1})^t \xi) \\ &= \sigma_j^{(2)}(\phi(x), (d\phi^{-1})^t \xi). \end{aligned}$$

Now Proposition 3.15 can be reformulated as follows:

(10.70) PROPOSITION: *Let* $\phi : U_1 \to U_2$ *denote an affine map. Let* $\mathcal{V}_2 \subset TU_2$ *denote a hyperplane bundle and* $\{X_j^{(2)}, j = 0, 1, \ldots, d\}$ *a* $\mathcal{V}_2$-*frame of* $TU_2$. *Let* $\mathcal{V}_1$ *and* $\{X_j^{(1)}, j = 0, 1, \ldots, d\}$ *denote their respective pullbacks to* $TU_1$. *Finally, let* $\mathcal{F}^m(U_j)$ *and* $S_{\mathcal{V}_j}^m(U_j), \quad j = 1, 2$, *denote the respective symbol classes. Then* $\phi$ *induces a 1-1, onto map, the pullback* $\phi_{\mathcal{V}_2}$,

$$(10.71) \qquad \phi_{\mathcal{V}_2} : S_{\mathcal{V}_2}^m(U_2) \to S_{\mathcal{V}_1}^m(U_1)$$

*as follows:*

$$(10.72) \qquad \begin{aligned} \phi_{\mathcal{V}_2} q_2(x,\xi) &= q_2(\phi(x),(d\phi^{-1})^t\xi) \\ &= q_1(x,\xi) \in S^m_{\mathcal{V}_1}(U_1). \end{aligned}$$

*Therefore we also have the pullback:*

$$(10.73) \qquad \phi_{\mathcal{V}_2} : \mathcal{F}^m(U_2) \to \mathcal{F}^m(U_1),$$

*given by*

$$(10.74) \qquad \phi_{\mathcal{V}_2} f_2(x,\xi) = f_2(\phi(x),\xi) = f_1(x,\xi),$$

*since*

$$(10.75) \qquad \sigma^{(1)}(x,\xi) = \sigma_0^{(2)}(\phi(x),(d\phi^{-1})^t\xi).$$

(10.76) REMARK: We have already had occasion to use and we shall have further need of the so called *"Peetre inequalities"*:

Let $\xi, \eta \in \mathbf{R}^{d+1}$. Then

$$(1\text{-}.77) \qquad \frac{1}{2}(1+|\eta|)^{-1} \le \frac{1+|\xi+\eta|}{1+|\xi|} \le 2(1+|\eta|),$$

and

$$(10.78) \qquad \frac{1}{3}(1+\|\eta\|)^{-1} \le \frac{1+\|\xi+\eta\|}{1+\|\xi\|} \le 3(1+\|\eta\|)$$

*are "Peetre's inequality" and the "parabolic Peetre inequality," respectively.*

(10.79) REMARK: $q(x,\xi) \in S^m_{\mathcal{V}}(U)$ implies that there is a $\tilde{q}(x,\xi) \in S^m_{\mathcal{V}}(U)$, such that $\tilde{q} \sim q$ and $\tilde{q}(x,D)$ is properly supported. In fact $q(x,\xi) \in S^m_{\mathcal{V}}(U)$ implies that $q(x,\xi)$ is a $(\frac{1}{2},\frac{1}{2})$-symbol (Proposition 10.22). By Remark 9.67 (ii) there is a $(\frac{1}{2},\frac{1}{2})$-symbol $\tilde{q}(x,\xi)$, such that $\tilde{q}(x,D)$ is properly supported and

$$(10.80) \qquad \tilde{q} - q \in S^{-\infty}(U).$$

In view of the discussion leading up to Proposition 10.45 we have $S^{-\infty}(U) = S^{-\infty}_{\mathcal{V}}(U)$. Consequently

$$(10.81) \qquad \tilde{q} - q \in S^{-\infty}_{\mathcal{V}}(U).$$

Therefore

$$(10.82) \qquad \tilde{q} \in S^m_{\mathcal{V}}(U),$$

and

$$(10.83) \qquad \tilde{q} \sim q.$$

§11 *Group Structures on* $\mathbf{R}^{d+1}$

We continue to work in an open subset $U \subset \mathbf{R}^{d+1}$ with a hyperplane bundle $\mathcal{V} \subset TU$ and a $\mathcal{V}$-frame $X_j$, $j = 0, 1, \ldots, d$. Given a point $y \in U$, there is a unique choice of affine coordinates, $x_j$, $j = 0, 1, \ldots, d$, in $U$, such that $y$ is the origin and each $X_j$, $j = 0, 1, \ldots, d$, coincides with $\partial/\partial x_j$ at $y$. We refer to these as the *y-coordinates*. In these *y*-coordinates

$$(11.1) \qquad X_j = \frac{\partial}{\partial x_j} + \frac{1}{2} \sum_{k=0}^{d} \beta_{jk}(x) \frac{\partial}{\partial x_k}, \quad j = 0, 1, \ldots, d,$$

with $\beta_{jk}(0) = 0$. We define constants $b_{jk}$ (dependent on $y$) by

$$(11.2) \qquad b_{jk} = b_{jk}(y) = \left( \frac{\partial}{\partial x_k} \beta_{j0} \right)(0), \quad j, k = 1, \ldots, d.$$

These constants are used to define a binary composition on $\mathbf{R}^{d+1}$: $\mathbf{R}^{d+1} \times \mathbf{R}^{d+1} \to \mathbf{R}^{d+1}$, which is expressed in the *y*-coordinates as

$$(11.3) \qquad \begin{aligned} (x \cdot z)_0 &= x_0 + z_0 + \frac{1}{2} \sum_{j,k=1}^{d} b_{kj} x_j z_k \\ &= x_0 + z_0 + \frac{1}{2} z'^t b x', \end{aligned}$$

$$(11.4) \qquad (x \cdot z)_j = x_j + z_j, \quad j = 1, \ldots, d.$$

With this composition $\mathbf{R}^{d+1}$ is a group; it is abelian if and only if the matrix $b = (b_{jk})$ is symmetric, and is otherwise a two step nilpotent Lie group. We call this the *y-group structure*. We note that (1.23) implies that the *y*-group is abelian for every $y \in U$ if and only if the bundle $\mathcal{V}$ is integrable: $[X, Y] \in \mathcal{V}$ for every $X, Y \in \mathcal{V}$.

We say that an operator on functions defined on $\mathbf{R}^{d+1}$ is *y-invariant* if it commutes with the left-translations of the *y*-group structure. For each $j$ the unique *y*-invariant vector field which agrees with $\partial/\partial x_j$ (and therefore with $X_j$) at $y$ is the infinitesimal right-translation $X_j^y$:

$$(11.5) \qquad X_j^y f(x) = \frac{d}{dt} f(x \cdot te_j)|_{t=0},$$

where the $e_j$, $j = 0, 1, \ldots, d$, are the canonical basis vectors for the *y*-coordinates. Thus

$$(11.6) \qquad X_0^y = \frac{\partial}{\partial x_0},$$

$$(11.7) \qquad X_j^y = \frac{\partial}{\partial x_j} + \frac{1}{2} \sum_{k=1}^{d} b_{jk} x_k \frac{\partial}{\partial x_0}, \qquad j = 1, \dots, d.$$

The $y$-group structure (11.3)–(11.4) does not depend on the choice of $\mathcal{V}$-frame. More precisely we have

(11.8) PROPOSITION: *Different $\mathcal{V}$-frames produce isomorphic $y$-groups.*

Proof. We consider two separate cases.

(i) Let $\tilde{X}_j$, $j = 0, 1, \dots, d$ denote a new $\mathcal{V}$-frame, where

$$(11.9) \qquad \begin{cases} \tilde{X}_0 = X_0, \\ \tilde{X}' = a(u) X', \end{cases}$$

where $a(u)$, $u \in U$, is non-singular. Set $a(y) = a = (a_{ij})$. To calculate the $y$-coordinates, $\tilde{x}$, relative to the $\tilde{X}$-frame we write

$$(11.10) \qquad X_j = \frac{\partial}{\partial x_j} + \frac{1}{2} \sum_{k,l=0}^{d} d_{jkl} x_l \frac{\partial}{\partial x_k} + \cdots,$$

$j = 0, 1, \dots, d$, where

$$(11.11) \qquad d_{jkl} = \left( \frac{\partial}{\partial x_l} \beta_{jk} \right)(0),$$

$$(11.12) \qquad b_{jl} = d_{j0l}, \qquad j, l = 1, \dots, d.$$

At $y$

$$(11.13) \qquad \sum_{m=1}^{d} a_{jm} \frac{\partial}{\partial x_m} = \frac{\partial}{\partial \tilde{x}_j},$$

which yields the $y$-coordinates relative to the $\tilde{X}$-frame:

$$(11.14) \qquad x_0 = \tilde{x}_0, \qquad x' = a^t \tilde{x}'.$$

We substitute (11.14) into $\tilde{X}' = aX'$, where $X$ is given by (11.10). This yields the $y$-invariant vector fields relative to the $\tilde{X}$-frame:

$$(11.15) \qquad \tilde{X}_0^y = \frac{\partial}{\partial \tilde{x}_0},$$

$$(11.16) \qquad \tilde{X}_j^y = \frac{\partial}{\partial \tilde{x}_j} + \frac{1}{2} \sum_{k=1}^{d} (aba^t)_{jk} \tilde{x}_k \frac{\partial}{\partial \tilde{x}_0}, \qquad j = 1, \dots, d.$$

Therefore

$$(x \cdot z)_0 = x_0 + z_0 + \frac{1}{2} z'^t b x'$$

(11.17)
$$= x_0 + z_0 + \frac{1}{2} \tilde{z}'^t (aba^t) \tilde{x}'$$

$$= (\tilde{x} \cdot \tilde{z})_0,$$

and

(11.18)                          $$(x \cdot z)' = a^t (\tilde{x} \cdot z)'.$$

This shows that the change of $\mathcal{V}$-frames, (11.9) produces the coordinate change (11.14) in the $y$-group.

   (ii) Next we leave $X_1, \ldots, X_d$ fixed and change $\tilde{X}_0$, i.e.

(11.19)          $$\tilde{X}_0(u) = \sum_{j=0}^{d} q_j(u) X_j(u), \quad q_0(y) \neq 0,$$

(11.20)          $$\tilde{X}_j(u) = X_j(u), \quad j = 1, \ldots, d, \quad u \in U.$$

The $y$-coordinates, $\tilde{x}$, relative to the $\tilde{X}$-frame are given by

(11.21)          $$x_0 = q_0 \tilde{x}_0 \quad \text{and} \quad x_j = q_j \tilde{x}_0 + \tilde{x}_j, \quad j = 1, \ldots, d,$$

where we set $q_j(y) = q_j$, $j = 0, 1, \ldots, d$. The $y$-invariant vector fields are

(11.22)          $$\tilde{X}_0^y = \frac{\partial}{\partial \tilde{x}_0},$$

(11.23)          $$\tilde{X}_j^y = \frac{\partial}{\partial \tilde{x}_j} + \frac{1}{2} \sum_{k=1}^{d} \frac{b_{jk}}{q_0} \tilde{x}_k \frac{\partial}{\partial \tilde{x}_0}.$$

Consequently, the change of $\mathcal{V}$-frames, $X \to \tilde{X}$, introduces a coordinate change in the $y$-group:

(11.24)          $$(x_0, x') = (q_0 \tilde{x}_0, \tilde{x}'),$$

since

$$(x \cdot z)_0 = x_0 + z_0 + \frac{1}{2} z'^t b x'$$

(11.25)
$$= q_0 \left( \tilde{x}_0 + \tilde{z}_0 + \frac{1}{2} \tilde{z}'^t \frac{b}{q_0} \tilde{x}' \right)$$

$$= q_0 (\tilde{x} \cdot \tilde{z})_0,$$

and

$$(11.26) \qquad (x \cdot z)' = (\tilde{x} \cdot \tilde{z})'.$$

This proves Proposition 11.8.

In the original coordinates the $y$-group structure makes $y$ the identity element and depends smoothly on $y$ (since the structure constants $b_{jk}(y)$ and the affine transformation from $y$-coordinates to original coordinates all depend smoothly on $y$).

Let $\sigma_j$ and $\sigma_j^y$ denote the symbols of the operators $i^{-1}X_j$ and $i^{-1}X_j^y$, respectively. Thus, in the $y$-coordinates

$$(11.27) \qquad \sigma_j(x,\xi) = \xi_j + \frac{1}{2} \sum_{k=0}^{d} \beta_{jk}(x)\xi_k, \quad j = 0, 1, \ldots, d,$$

and

$$(11.28) \qquad \sigma_0^y(x,\xi) = \xi_0,$$

$$(11.29) \qquad \sigma_j^y(x,\xi) = \xi_j + \frac{1}{2} \sum_{k=1}^{d} b_{jk}x_k\xi_0, \quad j = 1, \ldots, d.$$

Because of the choice of the $b_{jk}$'s we have the following relations which will be important later:

$$(11.30) \qquad \sigma_j - \sigma_j^y = \sum_{k=0}^{d} \alpha_{jk}(x)\sigma_k^y, \quad j = 0, 1, \ldots, d,$$

where

$$(11.31) \qquad \alpha_{jk}(0) = 0, \quad j, k = 0, 1, \ldots, d,$$

and

$$(11.32) \qquad |\alpha_{j)}(x)| \leqq C(y)(|x_0| + |x'|^2), \quad j = 1, \ldots, d.$$

Here $C$ is locally bounded on $U$. These relations characterize the constants $b_{jk}$, $j, k = 1, \ldots, d$, and indeed the group structure is chosen precisely to be as close to abelian as is consistent with the approximation result (11.30)–(11.32) for the left-invariant vector fields corresponding to the given $X_j$, $j = 0, 1, \ldots, d$.

In the $y$-group structure, the left-translations are affine maps. Therefore the effect of conjugating a pseudodifferential operator by a left-translation is calculated as in Proposition 3.15. It follows that a pseudodifferential operator is $y$-invariant if and only if its symbol $q$ has the form:

(11.33)
$$q(x,\xi) = f(\sigma^y(x,\xi))$$
$$= f(\sigma_0^y(x,\xi),\ldots,\sigma_d^y(x,\xi)),$$

where $f$ is in $C^\infty(\mathbf{R}^{d+1})$.

If $q \in S_V^m(U)$, i.e.

(11.34)
$$q(x,\xi) = f(x,\sigma(x,\xi)), \quad f \in \mathcal{F}^m(U),$$

then for any $y \in U$ we may associate to the $\mathcal{V}$-operator $Q = q(x,D)$ a $y$-invariant pseudodifferential operator $Q^y$ with symbol $q^y$,

(11.35)
$$q^y(x,\xi) = f(y,\sigma^y(x,\xi)).$$

In particular, when $Q = X_j$ we obtain $X_j^y$ above. Note that the symbols of $Q$ and of the approximating $y$-invariant operator $Q^y$ coincide at $y$:

(11.36)
$$q^y(0,\xi) = f(y,\sigma^y(0,\xi)) = f(y,\sigma(0,\xi)) = q(0,\xi)$$

when given in $y$-coordinates, and that $q^y$ is the unique $y$-invariant symbol with this property.

There is a one-parameter family of dilations on $\mathbf{R}^{d+1}$ associated to the $y$-group structure. We define them in the $y$-coordinates by

(11.37)
$$\delta_\lambda u(x) = \delta_{\lambda,y}u(x) = u_\lambda(x) = u(\lambda \cdot x), \quad \lambda > 0,$$

where, as before, $\lambda \cdot x = (\lambda^2 x_0, \lambda x')$, $x \in \mathbf{R}^{d+1}$. These are automorphisms for the $y$-group structure.

A pseudodifferential operator $Q$ acting on $\mathcal{E}'(\mathbf{R}^{d+1})$ is said to be $y$-*homogeneous of degree $m$* if

(11.38)
$$\delta_\lambda^{-1}\Omega\delta_\lambda = \lambda^m Q, \quad \lambda > 0.$$

In particular, the vector field $X_0^y$ is $y$-homogeneous of degree two and for $j > 0$ the vector field $X_j^y$ is $y$-homogeneous of degree one. For a symbol of the form (11.33) the necessary and sufficient condition for $y$-homogeneity of degree $m$ of the corresponding operator is that the function $f$ belong to $\mathcal{F}_m$. (This is, in general, incompatible with the requirement that $f$ be smooth at $\sigma = 0$.)

A very useful corollary of Proposition 10.70 obtains when the affine transformation $\phi$ produces $y$-coordinates. Thus let

(11.39)
$$X_j = \sum_{k=0}^{d} \beta_{jk}(z)\frac{\partial}{\partial z_k}, \quad j = 0, 1, \ldots, d,$$

denote a $\mathcal{V}$-frame. We inroduce $y$-coordinates $x$ by

(11.40) $$z = y + \beta(y)^t x = \psi_y^{-1}(x).$$

Let

(11.41) $$q(z,\xi) = f(z, \sigma(z,\xi)).$$

Then, according to (10.74)

(11.42) $$\begin{aligned} f(y,\xi) &= f(\psi_y^{-1}(0),\xi) \\ &= \tilde{f}(0,\xi) \\ &= \tilde{q}(0,\xi), \end{aligned}$$

where $\tilde{q}(x,\xi) = \tilde{f}(x, \sigma(x,\xi))$ denotes $q(z,\xi)$ in $y$-coordinates. Thus we have derived

(11.43) PROPOSITION: Let $q(z,\xi) = f(z, \sigma(z,\xi))$ belong to the class $S_{m,\mathcal{V}}(U)$ or to $S_{\mathcal{V}}^m(U)$ and let $q(y;x,\xi)$ denote $q$ in $y$-coordinates. Then

(11.44) $$f(y,\xi) = q(y;0,\xi).$$

The converse is now self-evident.

(11.45) PROPOSITION: Let $Q$ denote a properly supported operator on $U$. Fix $y \in U$ and let $x$ stand for $y$-coordinates. Suppose that at $x = 0$ $Q$ has a "symbol" in the following sense:

(11.46) $$Qu(0) = \int q(y;0,\xi)\hat{u}(\xi)d\xi, \quad u \in C_c^\infty(U).$$

If

(11.47) $$f(y,\xi) = q(y;0,\xi) \in \mathcal{F}^m(U),$$

then $Q$ is a $\mathcal{V}$-operator with symbol

(11.48) $$q(z,\eta) = f(z), \sigma(z,\eta)).$$

§12 Composition of $y$-Invariant Operators

A key step in showing that the composition of $\mathcal{V}$-operators is a $\mathcal{V}$-operator is to obtain the result in the $y$-invariant case, i.e. when $\sigma(x,\xi) \equiv \sigma^y(x,\xi)$. For the time being we fix $y \in \mathcal{U}$ and use the $y$-coordinates on $\mathbf{R}^{d+1}$, so that

$\sigma^y(0,\xi) = \xi$. For the $y$-invariant case we may simplify the corresponding function spaces.

(12.1) DEFINITION: *For $k \in \mathbf{Z}$, $\mathcal{F}_k$ is the space of functions $f$ belonging to $C^\infty(\mathbf{R}^{d+1}\backslash 0)$ which are $H$-homogeneous of degree $k$ with respect to the dilations (10.4). For $m \in \mathbf{Z}$, $\mathcal{F}^m$ is the space of functions $f \in C^\infty(\mathbf{R}^{d+1})$ which have an asymptotic expansion:*

$$(12.2) \qquad\qquad f \sim \sum_{j=0}^{\infty} f_{m-j}, \quad f_k \in \mathcal{F}_k,$$

*in the sense that for all multi-orders $\alpha$ and all $N \in \mathbf{N}$,*

$$(12.3) \qquad\qquad \left| D_\xi^\alpha \left( f - \sum_{j<N} f_{m-j} \right) \right| \leq C_{\alpha N} \|\xi\|^{m-N-\langle\alpha\rangle}.$$

*Here again $\| \; \|$ is the $H$-homogeneous norm (10.8) and $\langle\alpha\rangle$ is the $H$-homogeneous weight:* $2\alpha_0 + (\alpha_1 + \cdots + \alpha_n)$.

(12.4) DEFINITION: *For $m \in \mathbf{Z}$, $S_y^m$ is the space of symbols $q$ belonging to $C^\infty(\mathbf{R}^{d+1} \times \mathbf{R}^{d+1})$ of the form*

$$(12.5) \qquad\qquad q(x,\xi) = f(\sigma^y(x,\xi)), \quad f \in \mathcal{F}^m,$$

*in the $y$-coordinates on $\mathbf{R}^{d+1}$. Thus in particular*

$$(12.6) \qquad\qquad q(0,\xi) = f(\xi).$$

*We say that $q$ is the symbol associated to $f$, and that $q(x,D)$ is the operator associated to $f$.*

(12.7) REMARKS: A function $f \in \mathcal{F}^m$ satisfies the estimates

$$(12.8) \qquad\qquad |D^\alpha f(\xi)| \leq C_\alpha (1 + \|\xi\|)^{m-\langle\alpha\rangle}.$$

Conversely such a function defines a symbol $q$ by (12.5) and a corresponding operator $q(x,D)$. We continue to speak of the associated symbol and associated operator.

(12.9) PROPOSITION: *Suppose $f \in C^\infty(\mathbf{R}^{d+1})$ satisfies the estimates (12.8). Then the associated operator $Q = q(x,D)$ maps $\mathcal{S}(\mathbf{R}^{d+1})$ to itself. Moreover, $Q = 0$ if and only if $f = 0$.*

Proof. Clearly

$$(12.10) \qquad\qquad |q(x,\xi)| \leq C(1 + |x|)^{|m|}(1 + |\xi|)^{|m|}.$$

Therefore the usual formula

$$(12.11) \qquad\qquad Qu(x) = \int e^{i\langle x,\xi\rangle} q(x,\xi)\hat{u}(\xi)d\xi$$

defines $Qu$ for $u \in \mathcal{S}$ and shows that $Qu$ is a continuous function of polynomial growth in $x$. To prove rapid decrease in $x$ we use the identity

$$(12.12) \qquad e^{i\langle x, \xi \rangle} = (1 + |x|^2)^{-N} (1 - \Delta_\xi)^N e^{i\langle x, \xi \rangle},$$

where $\Delta_\xi$ is the Laplacian in the $\xi$-variables. Inserting the right side of (12.12) into (12.11) and integrating by parts, we replace the integrand by one which is dominated by

$$(12.13) \qquad (1 + |x|)^{|m| - 2N} (1 + |\xi|)^{|m|}.$$

Derivatives of $Qu$ are estimated in the same way.

Suppose that $Q = 0$. Then (12.11) with $x = 0$ implies $f(\xi) = q(0, \xi)$ vanishes identically: $\hat{u}$ runs through $\mathcal{S}(\mathbf{R}^{d+1}) \supset C_c^\infty(\mathbf{R}^{d+1})$.

An obvious consequence of Proposition 12.9 is that the operators associated to symbols of type $S_y^m$ can be composed. As we shall see, the composition gives an operator of the same type. The first step in proving this result is to show that the composed operator is associated to a function satisfying the correct extimates.

(12.14) THEOREM: *Suppose $Q_j = q_j(x, D)$, where $q_j \in S_y^{m_j}$, $j = 1, 2$. Then the composition $Q = Q_1 Q_2$ is the operator associated to a function $f$ which satisfies the estimates (12.8) with $m = m_1 + m_2$.*

Since the proof of this theorem is somewhat intricate, we begin by outlining the argument. It is necessary to show that for all $u \in \mathcal{S}(\mathbf{R}^{d+1})$

$$(12.15) \qquad Qu(0) = \int f(\xi) \hat{u}(\xi) \,\bar{d}\xi,$$

where $f$ satisfies (12.8). Conversely, it is sufficient to prove this: the operator associated to $f$ is $y$-invariant, and so is $Q$, so it is enough to show that they coincide at $x = 0$ (see (5.14)).

Next, note that it is easy to find a formal expression for the desired function $f$:

$$(12.16) \qquad \begin{aligned} Qu(0) &= \int q_1(0, \eta) [Q_2 u]\widehat{\phantom{m}}(\eta) \,\bar{d}\eta \\ &= \iint q_1(0, \eta) e^{-i\langle z, \eta \rangle} Q_2 u(z) \, dz \,\bar{d}\eta \\ &= \iiint q_1(0, \eta) e^{-i\langle z, \eta - \xi \rangle} q_2(z, \xi) \hat{u}(\xi) \,\bar{d}\xi \, dz \,\bar{d}\eta \\ &= \int f(\xi) \hat{u}(\xi) \,\bar{d}\xi, \end{aligned}$$

where, after one more change of variable,

$$f(\xi) = \iint e^{-i\langle z,\eta\rangle} q_1(0,\xi+\eta) q_2(z,\xi) dz \, d\eta$$

(12.17)

$$= \iint e^{-i\langle z,\eta\rangle} f_1(\xi+\eta) f_2(\sigma^y(z,\xi)) dz \, d\eta.$$

To make the integrals absolutely convergent we may introduce a smooth cut-off into (12.16):

(12.18)        $\phi_\varepsilon(z,\xi\eta) = \phi(\varepsilon[1 + |z|^2 + \|\eta\|^4 + \|\sigma^y(z,\xi)\|^4]),$

where $\phi \in C_c^\infty(\mathbf{R})$ and $\phi \equiv 1$ near the origin. Then we can interpret

(12.19)   $f(\xi) = \lim_{\varepsilon \searrow 0} \iint e^{-i\langle z,\eta\rangle} \phi_\varepsilon(z,\xi,\xi+\eta) f_1(\xi+\eta) f_2(\sigma^y(z,\xi)) dx \, d\eta,$

provided the limit can be shown to exist.

To show existence of (12.19) and to estimate it, we argue as follows. Suppose $L$ is a differential operator, $L = L(z,\xi,\eta,D_z,D_\eta)$, such that

(12.20)                $L^N(e^{-i\langle z,\eta\rangle}) = e^{-i\langle z,\eta\rangle}, \quad N \in \mathbf{N}.$

Then integration by parts converts the integral (12.19) to one with integrand

(12.21)        $e^{-i\langle z,\eta\rangle}(L^t)^N \{\phi_\varepsilon(z,\xi,\xi+\eta) f_1(\xi+\eta) f_2(\sigma^y(z,\xi))\},$

where $L^t$ is the transpose of $L$. If we are fortunate the integrand (12.21) will be absolutely integrable with estimates uniform in $\varepsilon$, $0 < \varepsilon \leq 1$. Since derivatives of $\phi_\varepsilon$ bring powers of $\varepsilon$, we can conclude both that the limit (12.19) exists and that it is given by integrating

(12.22)            $e^{-i\langle z,\eta\rangle}(L^t)^N \{f_1(\xi+\eta) f_2(\sigma^y(z,\xi))\}.$

We follow this course (with a technical modification), but for brevity we omit the use of $\phi_\varepsilon$ and proceed directly from (12.17) and (12.20) to (12.22). The technical modification is to use a smooth partition of unity to treat three separate parts of the region of integration and to use three separate operators $L$.

We now proceed to carry out the details. Choose $\phi \in C_c^\infty(\mathbf{R})$ with $\phi(t) = 1$ for $|t| \leq 1/9$ and $\phi(t) = 0$ for $|t| \geq 1/4$, and set

(12.23)        $\phi_1(\xi,\eta) = \phi(|\eta|^2[1 + |\xi|]^{-2}), \quad \phi_2(\xi,\eta) = 1 - \phi_1(\xi,\eta).$

Multiplying the integrand in (12.17) by $\phi_2$, we have reduced the region of integration to

(12.24)
$$|\eta| \geq \frac{1}{3}(1 + |\xi|).$$

On this region we use the operator

(12.25)
$$L = -(1 + |x|^2)^{-1}(1 - \Delta_\eta)(|\eta|^{-2}\Delta_z).$$

Then (12.20) holds, and the part of the integrand we are now considering becomes

(12.26)
$$e^{-i\langle z, \eta \rangle}(L^t)^N \{f_1(\xi + \eta)f_2(\sigma^y(z, \xi))\phi_2(\xi, \eta)\}.$$

Before differentiating, the integrand in (12.17) c n be estimated by

(12.27)
$$C(1 + \|\xi + \eta\|)^{m_1}(1 + \|\sigma^y(z, \xi)\|)^{m_2}.$$

Differentiation with respect to $\eta$ improves this estimate. As in the proof of Proposition 10.22, differentiation with respect to $z$ can harm the estimate at most by a factor of order

(12.28)
$$(1 + \|\sigma^y(z, \xi)\|)^{-1}(1 + |\xi|).$$

Now $\sigma^y(z, \xi) = \xi + a(z, \xi_0)$, where $a$ is bilinear, so

(12.29)
$$|\xi| \leq |\sigma^y(z, \xi)| + C|z|\,|\xi_0| \leq C_1(1 + |z|)(1 + \|\sigma^y(z, \xi)\|)^2.$$

Therefore we may dominate (12.28) by

(12.30)
$$(1 + |z|)^{\frac{1}{2}}(1 + |\xi|)^{\frac{1}{2}}.$$

Set

(12.31)
$$\langle z \rangle = 1 + |z|, \quad [\xi] = 1 + \|\xi\|, \quad \text{etc.}$$

We note that derivatives of $\phi_2(\xi, \eta)$ are bounded, uniformly in $\xi, \eta \in \mathbf{R}^{d+1}$. Then the preceding considerations show that (12.26) is dominated by

(12.32)
$$\langle z \rangle^{-N}|\eta|^{-2N}\langle \xi \rangle^N [\xi + \eta]^{m_1}[\sigma]^{m_2}, \quad \sigma = \sigma^y(z, \xi).$$

Since the integrand is supported in the region (12.24), we can replace (12.32) by

(12.33)
$$\langle z \rangle^{-N}\langle \eta \rangle^{-N}\langle \eta \rangle^{2|m_1|}\langle \eta \rangle^{|m_2|}\langle z \rangle^{|m_2|}$$
$$\leq C_{NM}\langle z \rangle^{|m_2|-N}\langle \eta \rangle^{m+2|m_1|+|m_2|-N}\langle \xi \rangle^{-M},$$

where we also used (10.78) and the bilinearity of $\sigma^y(z, \xi)$. For any given $M$ we may choose $N$ so large that (12.33) is integrable; the corresponding integral of (12.26) is dominated by $\langle \xi \rangle^{-M}$.

Next we consider the integrand in (12.17) multiplied by $\phi_1$. The result is supported where

$$(12.34) \qquad |\eta| \leq \frac{1}{2}(1 + |\xi|), \quad \text{so} \quad \langle \xi + \eta \rangle \sim \langle \xi \rangle.$$

We introduce a smooth cut-off in $z$ to subdivide into regions where

$$(12.35) \qquad\qquad\qquad |z| \leq 2;$$

$$(12.36) \qquad\qquad\qquad |z| \geq 1.$$

In the region (12.34), (12.36) we use the operator

$$(12.37) \qquad\qquad\qquad L = -|z|^{-2}\Delta_\eta$$

and argue as above. The corresponding integrand is replaced by one which is dominated by

$$(12.38) \qquad\qquad |z|^{-2N}[\xi + \eta]^{m_1 - 2N}[\sigma]^{m_2}.$$

On the region (12.34) we have

$$(12.39) \qquad \langle \xi \rangle^{\frac{1}{2}} \leq C \langle \xi + \eta \rangle^{\frac{1}{2}} \leq C_1[\xi + \eta].$$

Also

$$(12.40) \qquad\qquad [\sigma]^{m_2} \leq C \langle z \rangle^{|m_2|} \langle \xi \rangle^{|m_2|}$$

is self-evident. Consequently (12.38) is dominated by

$$(12.41) \qquad |z|^{|m_2| - 2N} \langle \xi \rangle^{|m_1|/2 + |m_2| + M - N} \langle \xi \rangle^{-M}.$$

Suppose $N$ is so large that the exponent in the second factor of (12.41) is negative. In that case, according to (12.34), we can replace $\xi$ by $\eta$ in the middle factor of (12.41). Therefore, we can again choose $N$ so as to dominate the corresponding integral by $\langle \xi \rangle^{-M}$.

Incorporating the remaining cut-off in $z$ into the symbol $q_2(z, \xi)$, we consider the remaining region (12.34), (12.35). The argument is similar but the estimates are subtler. Here we take

$$(12.42) \qquad L = l^{-1}[1 - \langle \xi \rangle^{-1}\Delta_z - \langle \xi \rangle \Delta_\eta],$$

where

$$(12.43) \qquad l = l(z, \xi, \eta) = 1 + \langle \xi \rangle^{-1}|\eta|^2 + \langle \xi \rangle |z|^2.$$

The integrand to be considered is

$$(12.44) \qquad e^{-i\langle z, \eta \rangle}(L^t)^N \{q_1(0, \xi + \eta) q_2(z, \xi)\phi_1(\xi, \eta)\}.$$

In the region (12.34), (12.35) we have

(12.45) $$\langle \xi \rangle \le C \langle \xi + \eta \rangle \le C_1 [\xi + \eta]^2,$$

(12.46) $$\langle \xi \rangle \le \langle \sigma \rangle + \langle \xi - \sigma \rangle \le C_1(\langle \sigma \rangle + |\sigma_0|) \le C_2 [\sigma]^2,$$

where $\sigma = \sigma^y(z, \xi)$. Therefore,

(12.47) $$|D_\eta^\alpha q_1(0, \xi + \eta)| \le C_\alpha [\xi + \eta]^{m_1 - |\alpha|} \le C_\alpha' [\xi + \eta]^{m_1} \langle \xi \rangle^{-|\alpha|/2},$$

(12.48) $$|D_z^\alpha q_2(z, \xi)| \le C_\alpha [\sigma]^{m_2 - |\alpha|} \langle \xi \rangle^{|\alpha|} \le C_\alpha' [\sigma]^{m_2} \langle \xi \rangle^{|\alpha|/2},$$

(12.49) $$|\langle \xi \rangle^{\frac{1}{2}} \nabla_\eta l| \le C l^{\frac{1}{2}} \le C l,$$

(12.50) $$|\langle \xi \rangle^{-\frac{1}{2}} \nabla_z l| \le C l^{\frac{1}{2}} \le C l.$$

It follows from (12.47)–(12.50) that (12.44) is dominated by

(12.51) $$[\xi + \eta]^{m_1} [\sigma]^{m_2} l^{-N} = [\xi]^m ([\xi + \eta][\xi]^{-1})^{m_1} ([\sigma][\xi]^{-1})^{m_2} l^{-N}.$$

Next we note that

(12.52) $$[\xi + \eta] \le [\xi] + [\eta] \le C l^{\frac{1}{2}} [\xi],$$

where we used

(12.53) $$[\eta] \le 1 + |\eta| \le 1 + [\xi] \langle \xi \rangle^{-\frac{1}{2}} |\eta|$$
$$\le [\xi](1 + \langle \xi \rangle^{-\frac{1}{2}} |\eta|) \le [\xi] l^{\frac{1}{2}}.$$

A similar argument, that uses (12.45), yields

(12.54) $$[\xi] \le [\xi + \eta] + [\eta] \le C[\xi + \eta] l^{\frac{1}{2}}.$$

Consequently, for all $m_1$, positive or negative, we have

(12.55) $$([\xi + \eta][\xi]^{-1})^{m_1} \le C l^{\frac{1}{2}|m_1|}.$$

As for $([\sigma][\xi]^{-1})^{m_2}$, we have

(12.56) $$[\sigma] \le [\xi] + [\sigma - \xi],$$

and

(12.57) $$[\sigma - \xi] \le 1 + |\sigma' - \xi'| \le 1 + C|z'||\xi_0|$$
$$\le C(1 + |z|\langle \xi \rangle^{\frac{1}{2}} \langle \xi \rangle^{\frac{1}{2}}) \le C[\xi] l^{\frac{1}{2}}.$$

Hence

(12.58) $$[\sigma] \le C[\xi] l^{\frac{1}{2}}.$$

Conversely

(12.59)                          $[\xi] \le [\sigma] + [\xi - \sigma],$

and from (12.57)

(12.60)                          $[\sigma - \xi] \le C\langle\xi\rangle^{\frac{1}{2}} l^{\frac{1}{2}} \le C[\sigma] l^{\frac{1}{2}}.$

Here we made use of (12.46). Therefore

(12.61)                          $[\xi] \le C[\sigma] l^{\frac{1}{2}},$

and

(12.62)                          $([\sigma][\xi]^{-1})^{m_2} \le C l^{\frac{1}{2}|m_2|},$

for all $m_2$. Therefore, for any $m_1$ and $m_2$ we can choose $2N > |m_1| + |m_2|$ and dominate (12.44) by

(12.63)                          $[\xi]^m l^{\frac{1}{2}(|m_1|+|m_2|-2N)}.$

Taking $N$ large, we may integrate (12.44) and find that the integral is dominated by $[\xi]^m$.

At this point we have shown that the operator $Q = Q_1 Q_2$ is defined on $C_c^{\infty}(\mathbf{R}^{d+1})$ and has symbol (12.5), where $f$ is dominated by $[\xi]^m = (1 + \|\xi\|)^m$. To complete the proof we must estimate derivatives of $f$. The argument is essentially the same as before. Differentiating once the formal expression (12.17) with respect to $\xi_j$, $j = 1, \ldots, d$ leads to an integrand which is a sum of terms each dominated by

(12.64)                $[\xi + \eta]^{m_1-1}[\sigma]^{m_2} + [\xi + \eta]^{m_1}[\sigma]^{m_2-1}.$

These may be treated exactly as before. When differentiating with respect to $\xi_0$ we need a better estimate. Here the corresponding integrand in (12.17) contains terms dominated by

(12.65)                $[\xi + \eta]^{m_1-2}[\sigma]^{m_2} + [\xi + \eta]^{m_1}[\sigma]^{m_2-2};$

however $(\sigma^y)_j$ is also dependent on $\xi_0$ for $j > 0$, and these terms are dominated by

(12.66)                $[\xi + \eta]^{m_1}[\sigma]^{m_2-1}|z|.$

In fact these terms are linear in $z$. Since

(12.67)                $z_j e^{-i\langle z, \eta\rangle} = i\frac{\partial}{\partial \eta_j}(e^{-i\langle z, \eta\rangle}),$

we may (formally) integrate by parts to get rid of the $z_j$'s and introduce $\eta$ derivatives. This converts the estimate (12.66) to

(12.68) $$[\xi + \eta]^{m_1 - 1} [\sigma]^{m_2 - 1}.$$

Once again, the resulting formal integrand may be transformed to give the desired estimate. Higher derivatives are handled in exactly the same way, so the proof of Theorem 12.14 is complete.

Our next goal is to sharpen Theorem 12.14 by showing that $f$ belongs to $\mathcal{F}^m$. To motivate the procedure, we suppose that $q_j$ has a single term, of degree $m_j$, in its asymptotic expansion. We sould expect that the $Q_j$ are "nearly homogeneous" in some sense, and that $Q$ will also be nearly homogeneous. To make this precise we start with the defining functions.

(12.69) DEFINITION: *A function $f \in C^\infty(\mathbf{R}^{d+1})$ is almost homogeneous of degree $m \in \mathbf{Z}$ if*

(12.70) $$\lambda^{-m} \delta_\lambda f - f \in \mathcal{S}(\mathbf{R}^{d+1}) \quad \text{for all} \quad \lambda > 0,$$

*where*

(12.71) $$\delta_\lambda f(\xi) = f(\lambda^2 \xi_0, \lambda \xi').$$

(12.72) PROPOSITION: *Suppose $f \in C^\infty(\mathbf{R}^{d+1})$. The following are equivalent:*

(a) *$f$ is almost homogeneous of degree $m$.*
(b) *$f \in \mathcal{F}^m$ and $f$ has a single term, of degree $m$, in its asymptotic expansion.*
(c) *There is $g \in \mathcal{F}_m$ such that for any cut-off function $\chi \in C^\infty(\mathbf{R}^{d+1})$ with $\chi \equiv 0$ near the origin and $\chi \equiv 1$ at $\infty$,*

$$f - \chi g \in \mathcal{S}(\mathbf{R}^{d+1}).$$

*Moreover the function $g$ in (c) is unique, given by*

(12.73) $$g(\xi) = \lim_{\lambda \to \infty} \lambda^{-m} f(\lambda \cdot \xi).$$

*We call $g$ the homogeneous part of $f$.*

Proof. It is almost immediate from the definitions that (b) $\Longleftrightarrow$ (c) $\Longrightarrow$ (a), that $g$ in part (c) coincides with the term in the asymptotic expansion in part (b), and that $g$ is unique. To complete the proof, assume (a). Then for any fixed $\lambda > 0$ and any $N > 0$,

(12.74) $$|\lambda^{-m} f(\lambda \cdot \xi) - f(\xi)| \leq C_{\lambda, N} (1 + \|\xi\|)^{-N}, \quad \xi \in \mathbf{R}^{d+1}.$$

Given $\mu > 0$ we may replace $\xi$ by $\mu \cdot \xi$ and conclude that for any $\xi \in \mathbf{R}^{d+1}\backslash 0$

$$(12.75) \qquad |(\lambda\mu)^{-m}f(\lambda\mu \cdot \xi) - \mu^{-m}f(\mu \cdot \xi)| \leq C_{\lambda,N}\mu^{-m-N}\|\xi\|^{-N}.$$

We define a sequence of functions $g_k$ on $\mathbf{R}^{d+1}\backslash 0$ by

$$(12.76) \qquad g_k(\xi) = (2^k)^{-m}f(2^k \cdot \xi), \quad k \in \mathbf{Z}_+.$$

Taking $\lambda = 2$ and $\mu = 2^k$ in (12.75) gives

$$(12.77) \qquad |g_{k+1}(\xi) - g_k(\xi)| \leq C_{2,N}2^{-k-1}\|\xi\|^{-N}$$

for large $N$. Therefore the $g_k$ converge uniformly on compact sets in $\mathbf{R}^{d+1}\backslash 0$ to a continuous function $g$. Moreover

$$(12.78) \qquad |g(\xi) - g_k(\xi)| \leq C_{2,N}2^{-k}\|\xi\|^{-N}.$$

Similar estimates apply to derivatives and show that $g$ is smooth. If we take $\mu = 2^k$ in (12.75) and let $k \to \infty$ with $\lambda$ fixed, we obtain homogeneity of $g$. Finally, with $k = 0$, (12.78) and the analogous estimates for derivatives show that $f \sim g$.

(12.79) DEFINITION: *If* $Q : C_c^{\infty}(\mathbf{R}^{d+1}) \to C^{\infty}(\mathbf{R}^{d+1})$ *and* $\lambda > 0$, *then* $Q^{(\lambda)}$ *is the operator*

$$(12.80) \qquad Q^{(\lambda)} = \delta_{\lambda}^{-1}Q\delta_{\lambda}.$$

(12.81) REMARK: An easy calculation shows that if $Q$ is the operator associated to a function $f$, then $Q^{(\lambda)}$ is the operator associated to $\delta_{\lambda}f$.

It is now a simple matter to sharpen Theorem 12.14.

(12.82) THEOREM: *Suppose* $Q_j = q_j(x, D)$, *where* $q_j \in S_y^{m_j}$, $j = 1, 2$. *Then* $Q = Q_1Q_2$ *is of the form* $Q = q(x, D)$, *where* $q \in S_y^m$, $m = m_1 + m_2$.

Proof. Suppose first that each $q_j$ has a single term in its asymptotic expansion, i.e. $q_j(0, \xi) = f_j(\xi)$, where $f_j \in \mathcal{F}^{m_j}$ has a single term $g_j$ in its asymptotic expansion. Let $f$ be the function of Theorem 12.14. According to Proposition 12.72, it is enough to prove that $f$ is almost homogeneous of degree $m$. By Remark 12.81 $\lambda^{-m}\delta_{\lambda}f - f$ is associated to the operator $\lambda^{-m}Q^{(\lambda)} - Q$. Now
(12.83)

$$\lambda^{-m}Q^{(\lambda)} - Q = \lambda^{-m_1}Q_1^{(\lambda)}\lambda^{-m_2}Q_2^{(\lambda)} - Q_1Q_2$$
$$= \lambda^{-m_1}Q_1^{(\lambda)}[\lambda^{-m_2}Q_2^{(\lambda)} - Q_2] + [\lambda^{-m_1}Q_1^{(\lambda)} - Q_1]Q_2.$$

By Remark 12.81, our assumption on $f_j$, and Proposition 12.72, the terms in brackets in (12.83) are associated to functions that belong to the space

$$(12.84) \qquad \mathcal{F}^{-\infty} = \bigcap_m \mathcal{F}^m = \mathcal{S}(\mathbf{R}^{d+1}).$$

By Theorem 12.14 and the fact that an operator determines its symbol (Proposition 12.9) it follows that $\lambda^{-m}\delta_\lambda f - f$ belongs to $\mathcal{S}(\mathbf{R}^{d+1})$, as desired.

Now we pass to the general case. We may choose $f_{j,k} \in \mathcal{F}^k$, $k \le m_j$, $j = 1,2$, so that each $f_{j,k}$ has a single term $g_{j,k} \in \mathcal{F}_k$ in its asymptotic expansion, and so that

$$(12.85) \qquad f_j \sim \sum_{k=0}^{\infty} f_{j,m_j-k}, \text{ as in (9.38).}$$

Let $Q_{j,k}$ be the operator associated to $f_{j,k}$. By what we have just proved, the composition $Q_{1,k}Q_{2,l}$ is associated to $f^{k,l} \in \mathcal{F}^{k+l}$ which has a single term $g^{k,l} \in \mathcal{F}_{k+l}$ in its asymptotic expansion. It follows from Theorem 12.14 that

$$(12.86) \qquad f - \sum_{k+l+N>0} f^{k,l} = O(\|\xi\|^{-N})$$

as $\|\xi\| \to \infty$, and similarly for derivatives. Therefore

$$(12.87) \qquad f \sim \sum_{j=0}^{\infty} \left\{ \sum_{k+l=m_1+m_2-j} g^{k,l} \right\}.$$

This proves Theorem 12.82.

(12.88) REMARK: For applications it is useful to restate Theorem 8.13 for symbols. Thus we denote by

$$(12.89) \qquad X_0^y = \frac{\partial}{\partial x_0},$$

$$(12.90) \qquad X_j^y = \frac{\partial}{\partial x_j} + \frac{1}{2} \sum_{k=1}^{d} b_{jk} x_k \frac{\partial}{\partial x_0}, \quad j = 1, \ldots, d,$$

the left-invariant vector fields in general $y$-coordinates. As in §8 we set $c_{jk} = \frac{1}{2}(b_{jk} - b_{kj})$, $s_{jk} = \frac{1}{2}(b_{jk} + b_{kj})$, $j, k = 1, \ldots, d$ and write

$$(12.91) \qquad X_j^y = \frac{\partial}{\partial x_j} + \frac{1}{2} \sum_{k=1}^{d} c_{jk} x_k \frac{\partial}{\partial x_0} + \frac{1}{2} \sum_{k=1}^{d} s_{jk} x_k \frac{\partial}{\partial x_0},$$

$j = 1, \ldots, d$. A quadratic change of variables,

$$(12.92) \qquad u = \chi(x) = \left( x_0 - \frac{1}{4} \sum_{j,k=1}^{d} s_{jk} x_j x_k, \ x' \right),$$

reduces the $y$-invariant vector fields, $X_j^y$, $j = 0, 1, \ldots, d$, to skew-symmetric form:

$$(12.93) \qquad X_0^y = \frac{\partial}{\partial u_0},$$

$$(12.94) \qquad X_j^y = \frac{\partial}{\partial u_j} + \frac{1}{2} \sum_{k=1}^{d} c_{jk} u_k \frac{\partial}{\partial u_0}, \quad j = 1, \ldots, d.$$

Now the Fourier transform of (8.14) yields

(12.95) THEOREM: *Let $Q$ be a $y$-operator with $q(x, \xi) = f(\sigma^y(x, \xi)) \in S_y^m$ its symbol in $y$-coordinates and $q'(u, \xi) = f'(\sigma'(u, \xi))$ its symbol in skew-symmetric $y$-coordinates, where $u = \chi(x)$ is given by (12.92). Then*

$$(12.96) \qquad f(\sigma) = \iint e^{i\langle z, \gamma \rangle - i \langle \chi(z), \sigma \rangle} f'(\gamma) dz \, d\gamma,$$

*and*

$$(12.97) \qquad f'(\gamma) = \iint e^{i\langle z, \sigma \rangle + i \langle \chi(z), \gamma \rangle} f(\sigma) dz \, d\sigma.$$

§13  *The #-Composition of Homogeneous Symbols*

Let $f_j \in \mathcal{F}^{m_j}$ with $f_j \sim g_j \in \mathcal{F}_{m_j}$, $j = 1, 2$. Using (12.17) we set

$$(13.1) \qquad \begin{aligned} f(\xi) &= f_1 \#_y f_2(\xi) \\ &= \iint e^{-i\langle z, \eta \rangle} f_1(\xi + \eta) f_2(\sigma^y(z, \xi)) dz \, d\eta. \end{aligned}$$

The integral is defined by (12.19). According to the first part of the proof of Theorem 12.82 $f \sim g \in \mathcal{F}_m$, $m = m_1 + m_2$. It is clear that $g$ is uniquely determined by $g_j$, $j = 1, 2$, and that this composition makes $\bigcup_m \mathcal{F}_m$ an algebra. The composition depends on $y$ and we write

$$(13.2) \qquad g = g_1 \#_y g_2.$$

We shall need to investigate the $y$-dependence of the map (13.2). We topologize $\mathcal{F}_m$ by identifying it, via restriction, with $C^\infty(\{\xi \in \mathbf{R}^{d+1} : \|\xi\| = 1\})$.

(13.3) PROPOSITION: *Given $g_j \in \mathcal{F}_{m_j}$, $j = 1, 2$, the map $y \to g_1 \#_y g_2$ is a smooth map from $U$ to $\mathcal{F}_m$, $m = m_1 + m_2$.*

Proof. Choose $f_j \in \mathcal{F}^{m_j}$ with $f_j \sim g_j$, $j = 1, 2$, and set $f = f_1 \#_y f_2$. Let us topologize $\mathcal{F}^m$ by using the seminorms

$$(13.4) \qquad N_\alpha(f) = \sup_\xi (1 + \|\xi\|)^{\langle \alpha \rangle - m} |D^\alpha f(\xi)|.$$

The mapping from an almost homogeneous function $f$ to the corresponding homogeneous function $g$ is clearly continuous from the $\mathcal{F}^m$ topology to $\mathcal{F}_m$. In fact the topology of $\mathcal{F}_m$ can be defined by the seminorms.

$$N_{h,\alpha}(g) = \limsup_{R \to \infty} \left[ \max_{\|\xi\| = R} (1 + \|\xi\|)^{\langle \alpha \rangle - m} |D^\alpha g(\xi)| \right],$$

whence the continuity of the map $f \to g$, $f$ almost homogeneous, follows immediately. Therefore it is enough to show that the map $y \to f_1 \#_y f_2$ is smooth from $U$ to $\mathcal{F}^m$. This we do by using (13.1), i.e.

$$(13.5) \qquad f_1 \#_y f_2(\xi) = \iint e^{-i\langle z, \eta \rangle} f_1(\xi + \eta) f_2(\sigma^y(z, \xi)) dz \, \bar{d}\eta,$$

which is given meaning as in (12.19). Looking at the various convergent integrals in the proof of Theorem 12.14 which represent pieces of (13.5) after formal integrations by parts, one sees immediately that our map is continuous with respect to $y$. Now

$$(13.6) \qquad \sigma^y(z, \xi) = \xi + a(y; z, \xi_0)$$

where $a$ is smooth and $a(y; \cdot, \cdot)$ is bilinear. Differentiating with respect to $y$ lowers the degree of homogeneity of $f_2$ by 1 but introduces the factor $\frac{\partial a}{\partial y}$ which has degree 2. However this factor is linear in $z$ so as above we may use (12.67) to replace $z_j$ by a differentiation which lowers the degree of $f_1$. As in the proof of Theorem 12.14 we can now obtain the $\mathcal{F}_m$-estimates for the first derivatives with respect to $y$. The argument can be iterated, yielding estimates for all derivatives.

We are now in a position to introduce corresponding compositions in the function spaces $\mathcal{F}_m(U)$ and $S_{m,\nu}(U)$.

(13.7) DEFINITION: *Given $g_j \in \mathcal{F}_{m_j}(U)$, $j = 1, 2$, we take $g_1 \# g_2$ to be the function defined on $U \times (\mathbf{R}^{d+1} \backslash 0)$ by*

$$(13.8) \qquad g_1 \# g_2(y, \cdot) = g_1(y, \cdot) \#_y g_2(y, \cdot).$$

(13.9) PROPOSITION: *$g_1 \# g_2$ belongs to $\mathcal{F}_{m_1 + m_2}(U)$.*

Proof. The maps $y \to g_j(y, \cdot)$ are smooth from $U$ to $\mathcal{F}_{m_j}$. Combining this with Proposition 13.3, we have the desired conclusion: $g_1 \# g_2$ belongs to $C^\infty(U \times \mathbf{R}^{d+1} \backslash 0))$.

(13.10) DEFINITION: *Suppose $q_j \in S_{m_j, \mathcal{V}}(U)$, $j = 1, 2$, with*

$$(13.11) \qquad q_j(z, \xi) = g_j(z, \sigma(z, \xi)), \quad g_j \in \mathcal{F}_{m_j}(U), \ j = 1, 2.$$

*Then $q_1 \# q_2 \in S_{m_1 + m_2, \mathcal{V}}(U)$ is the symbol defined by*

$$(13.12) \qquad q_1 \# q_2(z, \xi) = g_1 \# g_2(z, \sigma(z, \xi)).$$

The #-composition of homogeneous symbols, (13.12), as we shall show in the next section, produces precisely the principal symbol in the composition of homogeneous $\mathcal{V}$-operators. If we avail ourselves of the discussion leading up to Definition 13.10 we have the following explanation of (13.12):

Let $Q_j$ denote a properly supported operator corresponding to $q_j \in S_{m_j, \mathcal{V}}(U)$, in the sense that $Q_j = \tilde{q}_j(z, D)$, where $\tilde{q}_j \sim q_j$, $j = 1, 2$. Let

$$(13.13) \qquad \tilde{q}_j(z, \xi) = f_j(z, \sigma(z, \xi)),$$

and $f_j \sim g_j \in \mathcal{F}_{m_j}(U)$, $j = 1, 2$. Given $y \in U$, let $Q_j^y$ denote the $y$-invariant operator with symbol

$$(13.14) \qquad \tilde{q}_j^y(x, \xi) = f_j(y, \sigma^y(x, \xi)), \quad j = 1, 2,$$

where $x$ denotes $y$-coordinates. Then $Q_1^y Q_2^y$ is a $y$-invariant operator which has symbol $\tilde{q}_y(x, \xi)$, given by

$$(13.15) \qquad \tilde{q}_y(x, \xi) = f(y, \sigma^y(x, \xi)).$$

Now $f$ defines, uniquely, its homogeneous part. For fixed $y$ we have

$$(13.16) \qquad f(y, \xi) \sim (g_1(y, \cdot) \#_y g_2(y, \cdot))(\xi) \in \mathcal{F}_{m_1 + m_2}.$$

Consequently

$$(13.17) \qquad \begin{aligned} \tilde{q}_y(x, \xi) &\sim (g_1 \# g_2)(y, \sigma^y(x, \xi)) \\ &= (q_1 \# q_2)^y(x, \xi). \end{aligned}$$

Setting $x = 0$ we find

$$(13.18) \qquad \tilde{q}_y(0, \xi) \sim (g_1 \# g_2)(y, \xi),$$

which brings us full circle.

## §14 *The Composition of $V$-Operators*

We continue to assume that $U \subset \mathbf{R}^{d+1}$ is an open set with a given hyperplane bundle $V \subset TU$. Our goal in §14 is to derive a complete asymptotic expansion for the symbol of the composition of two $V$-operators, Theorem 14.7. It has the following simple corollary:

(14.1) THEOREM: *Suppose $Q_j$ is a properly supported $V$-operator of degree $m_j$, $j = 1, 2$. Then $Q = Q_1 Q_2$ is a properly supported $V$-operator of degree $m = m_1 + m_2$. Moreover, if $q_{j,m_j}$ denotes the principal symbol of $Q_j$, $j = 1, 2$, then the principal symbol $q_m \in S_{m,V}(U)$ of $Q$ is given by*

$$(14.2) \qquad q_m = q_{1,m_1} \# q_{2,m_2}.$$

Fix $y \in U$. To calculate the full asymptotic expansion of $q(y, \xi)$ we start by choosing $y$-coordinates, $x$, on $U$, so that $y$ is identified with $0 \in \mathbf{R}^{d+1}$ and $\sigma_j(0, \xi) = \xi_j$, $j = 0, 1, \ldots, d$. The terms of the asymptotic expansion will be expressed in the following notation.

(14.3) DEFINITION: *If $q(x, \xi) = f(x, \sigma(x, \xi))$, we set*

$$(14.4) \qquad q^{(\delta)}(x, \xi) = \partial_\sigma^\delta f(x, \sigma(x, \xi)),$$

$$(14.5) \qquad q_\alpha^{(\beta;\gamma)}(x, \xi) = D_x^\alpha \partial_\sigma^\beta f(x, \sigma(x, \xi))(\sigma(x, \xi))^\gamma.$$

*We also define the functions $e_{\beta\gamma}(x)$ by*

$$(14.6) \qquad [\sigma(x, \xi) - \sigma^0(x, \xi)]^\beta = \sum_{|\gamma|=|\beta|} e_{\beta\gamma}(x)(\sigma^0(x, \xi))^\gamma.$$

In the following theorem we continue to identify $y \in U$ with the origin in $\mathbf{R}^{d+1}$ via the $y$-coordinates.

(14.7) THEOREM: *Under the assumptions of Theorem 14.1 let $q_{j,s} \in S_{s,V}(U)$ denote the terms in the asymptotic expansion of $q_j$, $j = 1, 2$. Let $q(0, \xi) = q_1 \circ q_2(0, \xi)$ denote the "symbol" of $Q_1 Q_2$ at $x = 0$, in the sense of Proposition 11.45. Then $q_1 \circ q_2(0, \xi) \in \mathcal{F}^m$ and the term of order $r$ in its asymptotic expansion is*

$$(14.8) \quad (q_1 \circ q_2)_r(0, \xi) = \sum \frac{1}{\alpha! \beta! \delta!} (D^\delta e_{\beta\gamma})(0)[q_{1,s}]^{(\alpha+\delta)} \# [q_{2,t}]_\alpha^{(\beta;\gamma)}(0, \xi).$$

*The summation is restricted to indices which satisfy*

(14.9) $$r = s + t - \langle \alpha \rangle - \langle \delta \rangle - \langle \beta \rangle + \langle \gamma \rangle,$$

*and*

(14.10) $$-\langle \gamma \rangle + \langle \delta \rangle + \langle \beta \rangle \geq |\beta| = |\gamma|.$$

*Finally, as a function of $y$, $q(0,\xi) \in \mathcal{F}^m(U)$.*

(14.11) REMARK: The restrictions (14.9) and (14.10) imply that there are finitely many terms in the sum (14.8). Indeed

(14.12)
$$\begin{aligned}
m_1 \geq s &= r - t - \langle \gamma \rangle + \langle \alpha \rangle + \langle \delta \rangle + \langle \beta \rangle \\
&\geq r - t + (\langle \alpha \rangle + |\beta|) \geq r - m_2,
\end{aligned}$$

i.e.

(14.13) $$m_1 \geq s \geq r - m_2,$$

and similarly

(14.14) $$m_2 \geq t \geq r - m_1,$$

while

(14.15) $$\langle \alpha \rangle + |\beta| = \langle \alpha \rangle + |\gamma| \leq m - r,$$

and

(14.16) $$\langle \delta \rangle \leq m - r + 2|\gamma| \leq 3(m - r).$$

Now Proposition 11.45 yields the full asymptotic expansion for $q$.

(14.17) THEOREM: *Let $g_r \in \mathcal{F}_r(U)$ be defined by*

(14.18) $$g_r(y,\xi) = (q_1 \circ q_2)_r^y(0,\xi),$$

*where $(q_1 \circ q_2)_r^y(0,\xi) = (q_1 \circ q_2)_r(0,\xi)$ is given by (14.8). Set*

(14.19) $$q_r(y,\xi) = g_r(y, \sigma(y,\xi)).$$

*Then $q_r \in S_{r,\nu}(U)$ and*

(14.20) $$q \sim \sum_r q_r.$$

We begin the proof of Theorem 14.7 with a technical simplification. Since $Q_1 = q_1(x, D)$ is properly supported, and since we are working in a neighborhood of $x = 0$, we may replace $Q$ by the operator $Q_1 \phi Q_2$, where $\phi \in C_c^\infty(U)$ is $\equiv 1$ in a suitable neighborhood of the origin. Thus we may replace $q_2$ by $\phi q_2$ and assume that $q_2$ is compactly supported with respect to $x$.

The idea of the proof is quite simple: we use convenient Taylor expansions and the classical asymptotic expansion (for symbols of type $(\frac{1}{2}, \frac{1}{2})$ composed with classical symbols). The remainder terms are small, either by classical estimates or by the same process of estimation that we used in the proof of Theorem 12.14.

By assumption $q_2$ has the form

$$(14.21) \qquad q_2(x, \xi) = f_2(x, \sigma(x, \xi)), \quad f_2 \in \mathcal{F}^{m_2}(U).$$

The formal calculation leading to (12.17) also yields the formal expression

$$(14.22) \qquad q(0, \xi) = \iint e^{-i\langle x, \eta \rangle} q_1(0, \xi + \eta) f_2(x, \sigma(x, \xi)) dx \, d\eta.$$

We shall show that $q(0, \xi)$, defined by (14.22), belongs to $\mathcal{F}^m$ by explicitly calculating its defining asymptotic expansion. Let $M$ denote an arbitrary negative integer. Then

$$(14.23) \qquad q_j = \sum_{M < k \leq m_j} q_{j,k} + r_{j,M}^{(1)},$$

where $r_{j,M}^{(1)} \in S_{\mathcal{V}}^M(U)$. For the moment we neglect the remainder terms $r_{j,M}^{(1)}$, $j = 1, 2$. Thus we are left with composing a finite number of homogeneous terms; we may as well set

$$(14.24) \qquad q_j(x, \xi) = f_j(x, \sigma(x, \xi)),$$

$$(14.25) \qquad f_j \sim g_j \in \mathcal{F}_{m_j}(U), \quad j = 1, 2.$$

This we shall assume in what follows.

To calculate the composition of $q_1$ and $q_2$ we take the Taylor expansion of $f_2$ around $(0, \sigma^0(x, \xi))$:

$$(14.26) \qquad \begin{aligned} &f_2(x, \sigma(x, \xi)) \\ &= \sum_{|\alpha| + |\beta| < N} \frac{1}{\alpha! \beta!} \partial_x^\alpha \partial_\sigma^\beta f_2(0, \sigma^0(x, \xi)) x^\alpha (\sigma - \sigma^0)^\beta + r_N^{(2)}(x, \xi), \end{aligned}$$

where $\sigma = \sigma(x, \xi)$ and $\sigma^0 = \sigma^0(x, \xi)$. We rewrite (14.26) using (14.5) and (14.6). Associated to the symbol $[q_2]_\alpha^{(\beta;\gamma)}$ of (14.5) is the $x = 0$ invariant symbol

$$(14.27) \qquad [q_2]_\alpha^{(\beta;\gamma),0}(x, \xi) = D_x^\alpha \partial_\sigma^\beta f_2(0, \sigma^0(x, \xi))(\sigma^0(x, \xi))^\gamma.$$

Thus (14.26) becomes

$$f_2(x, \sigma(x,\xi))$$

(14.28)
$$= \sum_{|\alpha|+|\beta|<N} i^{|\alpha|} \sum_{|\gamma|=|\beta|} \frac{x^\alpha}{\alpha!\beta!} e_{\beta\gamma}(x)[q_2]_\alpha^{(\beta;\gamma),0}(x,\xi) + r_N^{(2)}(x,\xi).$$

To simplify the argument we consider a single term in the summation (14.28). The corresponding integral

(14.29)
$$\frac{1}{\alpha!\beta!} \iint e^{-i\langle x,\eta\rangle} [q_1]^{(\alpha),0}(0,\xi+\eta) e_{\beta\gamma}(x)[q_2]_\alpha^{(\beta;\gamma),0}(x,\xi) dx\, d\eta$$

(interpreted as in the proof of Theorem 12.14) is the symbol, at $x = 0$, of the operator

(14.30)
$$\frac{1}{\alpha!\beta!} [q_1]^{(\alpha),0}(x,D) E_{\beta\gamma} \circ [q_2]_\alpha^{(\beta;\gamma),0}(x,D),$$
$$= \frac{1}{\alpha!\beta!} Q_1^{(\alpha),0} E_{\beta\gamma} Q_{2,\alpha}^{(\beta;\gamma),0},$$

where we set

(14.31)
$$E_{\beta\gamma} u(x) = e_{\beta\gamma}(x)u(x).$$

We note that

(14.32)
$$q_j^0(x,\xi) = f_j(0, \sigma^0(x,\xi))$$

is the 0-invariant symbol associated to $q_j$, $j = 1,2$. Let us first compose $Q_1^{(\alpha),0}$ with $E_{\beta\gamma}$. At $x = 0$ its symbol is

(14.33)
$$\iint e^{-i\langle x,\eta\rangle} [q_1]^{(\alpha),0}(0,\xi+\eta) e_{\beta\gamma}(x) dx\, d\eta.$$

To calculate (14.33) we write out the Taylor expansion for $e_{\beta\gamma}(x)$:

(14.34)
$$e_{\beta\gamma}(x) = \sum_{|\delta|<K} \frac{1}{\delta!} (\partial^\delta e_{x'\beta\gamma})(0)x^\delta + \sum_{|\delta|=K} r_{K,\delta}^{(3)}(x)x^\delta,$$

where

(14.35)
$$r_{K,\delta}^{(3)}(x) = \frac{K}{\delta!} \int_0^1 (1-t)^{K-1} (\partial^\delta e_{x'\beta\gamma})(tx) dt.$$

Consequently, the symbol of (14.30) at $x = 0$ is

$$\frac{1}{\alpha!\beta!}\sum_{|\delta|<K}\frac{1}{\delta!}(D_x^\delta e_{\beta\gamma})(0)$$

(14.36)
$$\iint e^{-i\langle x,\eta\rangle}[q_1]^{(\alpha+\delta),0}(0,\xi+\eta)[q_2]_\alpha^{(\beta;\gamma),0}(x,\xi)dx\,\bar{d}\eta$$

$$+\frac{1}{\alpha!\beta!}\sum_{|\delta|=K}\iint e^{-i\langle x,\eta\rangle}[q_1]^{(\alpha+\delta),0}(0,\xi+\eta)$$

$$r_{K,\delta}^{(3)}(x)[q_2]_\alpha^{(\beta;\gamma),0}(x,\xi)dx\,\bar{d}\eta.$$

In view of the preceding, after neglecting remainder terms, we have shown that $q(0,\xi)$ is formally the sum of the symbols of the operators

(14.37)
$$\frac{1}{\alpha!\beta!\delta!}(D^\delta e_{\beta\gamma})(0)Q_1^{(\alpha+\delta),0}Q_{2,\alpha}^{(\beta;\gamma),0}$$

at $x = 0$. Since we assumed that $q_j$ has a single term of degree $m_j$, $j = 1,2$, in its asymptotic expansion, the symbol of (14.37), at $x = 0$, is asymptotic to

(14.38)
$$\frac{1}{\alpha!\beta!\delta!}(D^\alpha e_{\beta\gamma})(0)[q_{1,m_1}]^{(\alpha+\delta)}\#[q_{2,m_2}]_\alpha^{(\beta;\gamma)}(0,\xi),$$

and the symbol of (14.30) is asymptotic to

(14.39)
$$\sum_\delta\frac{1}{\alpha!\beta!\delta!}(D^\delta e_{\beta\gamma})(0)[q_{1,m_1}]^{(\alpha+\delta)}\#[q_{2,m_2}]_\alpha^{(\beta;\gamma)}(0,\xi)$$

at $x = 0$. Summing (14.39) over $\alpha$ and $\beta$ yields (14.20) when $q_1$ and $q_2$ are asymptotically homogeneous. In the general case one obtains the full asymptotic expansion (14.20) with $q_r(0,\xi)$ given by (14.8).

To complete the proof we must estimate the neglected remainder terms and show that they are negligible. We must also show that there are only a finite number of terms of a given degree in the expansion (14.8). We do the latter by establishing the limitation (14.10); (14.9) is self-explanatory.

(14.40) LEMMA: *The functions $e_{\beta\gamma}(x)$ of (14.6) satisfy the estimates*

(14.41)
$$e_{\beta\gamma}(x) = O(\|x\|^{|\beta|-\langle\beta\rangle+\langle\gamma\rangle})$$

*as $x \to 0$. Consequently*

(14.42)
$$D^\delta e_{\beta\gamma}(0) = 0$$

*whenever*

(14.43)
$$\langle\delta\rangle + \langle\beta\rangle - \langle\gamma\rangle < |\beta|.$$

Proof. The inequalities (11.32) imply (14.41) in the case $|\beta| = |\gamma| = 1$. The general (14.41) follows by taking products.

Clearly, (14.10) is a consequence of (14.42) and (14.43). It remains to estimate the neglected remainder terms, i.e. "error terms," coming from our various expansions. For this purpose we note that Theorem 12.14 holds under somewhat weaker assumptions. Let $f \in \mathcal{F}^m(U)$. Then

$$(14.44) \qquad |D_x^\alpha \partial_\xi^\beta f(x,\xi)| \le C_{\alpha,\beta}(1 + \|\xi\|)^{m-\langle\beta\rangle}$$

for all multi-indices $\alpha$ and $\beta$.

(14.45) PROPOSITION: *Let* $f_j(x,\xi) \in C^\infty(U \times \mathbf{R}^{d+1})$ *satisfy (14.44) with exponent* $m_j$, $j = 1, 2$. *Set*

$$(14.46) \qquad \sigma(x,\xi) = \xi + \frac{1}{2}\beta(x)\xi$$

*with* $\beta(0) = 0$, *where* $\sigma = (\sigma_0, \sigma_1, \ldots, \sigma_d)$, $\xi = (\xi_0, \xi_1, \ldots, \xi_d)$ *and* $\beta(x) = (\beta_{jk}(x))$. *Let* $\psi \in C_c^\infty(U)$ *with* $\operatorname{supp}\psi \subset \{x; |x| < 1\}$ *and assume that* $I + \beta(x)$ *is invertible when* $x \in \operatorname{supp}\psi$. *Then*

$$(14.47) \qquad \iint e^{-i\langle x,\eta\rangle} f_1(0, \xi + \eta)\psi(x)f_2(x, \sigma(x,\xi))dx\,d\eta$$

*satisfies (12.8) with* $m = m_1 + m_2$.

Proof. The first order Taylor expansion of $\beta_{jk}(x)$ yields

$$(14.48) \qquad \sigma_j(x,\xi) = \xi_j + \frac{1}{2}\sum_{k,l=0} \beta_{jkl}(x)x_k\xi_l$$

with

$$(14.49) \qquad |\beta_{jkl}(x)| \le \text{Const}.$$

for $j, k, l = 0, 1, \ldots, d$ and $x \in \operatorname{supp}\psi$. Consequently the proof of Theorem 12.14 also proves Proposition 14.45, if, in the argument, we replace the bilinearity of $\sigma^y(x,\xi)$ with (14.48) and (14.49).

We return to estimating the error terms in the proof of Theorem 14.7.
(i) $r_{j,M}^{(1)}$, the remainder in the asymptotic expansion of $q_j$, $j = 1, 2$, see (14.23), induces the following error terms:

$$(14.50) \qquad \iint e^{-i\langle x,\eta\rangle} q_1(0, \xi + \eta)r_{2,M}^{(1)}(x,\xi)dx\,d\eta,$$

$$(14.51) \qquad \iint e^{-i\langle x,\eta\rangle} r_{1,M}^{(1)}(0, \xi + \eta) \sum_{M<k\le m_2} q_{2,k}(x,\xi)dx\,d\eta, \text{ etc.},$$

$r_{j,M}^{(1)} \in S_v^M(U)$, $j = 1, 2$, where $M$ is an arbitrarily large negative integer. According to Proposition 14.45, (14.50) and (14.51) satisfy (12.8) with $m = m_1 + M$ and $m = M + m_2$, respectively, and, in view of the fact that $M$ can be chosen arbitrarily small, are negligible.

(ii) Next we estimate the error induced by neglecting the remainder term $r_N^{(2)}(x, \xi)$ of (14.26). The corresponding term in the formal integral (14.22) is the integral against $(1 - t)^{N-1}$ in the interval $0 \leq t \leq 1$ of a linear combination of terms, with $|\alpha| + |\beta| = N$, of the following form:

$$(14.52) \qquad \iint e^{-i\langle x, \eta \rangle} q_1(0, \xi + \eta)[f_2]_\alpha^{(\beta;0)}(tx, \sigma^t)x^\alpha[\sigma - \sigma^0]^\beta \, dx \, d\eta,$$

where

$$(14.53) \qquad \sigma^t = (1 - t)\sigma^0 + t\sigma, \ \sigma^0 = \sigma^0(x, \xi), \ \sigma = \sigma(x, \xi).$$

To estimate (14.52) we write

$$(14.54) \qquad [\sigma - \sigma^0]^\beta = \sum_{|\gamma|=|\beta|} e_{\beta\gamma}^t(x)(\sigma^t)^\gamma, \ e_{\beta\gamma}^0 = e_{\beta\gamma}.$$

From (11.10)

$$(14.55) \qquad \sigma_j - \sigma_j^0 = \left(\frac{1}{2}d_{j00}|x_0| + O(|x|^2)\right)\xi_0 + O(|x|)\xi,$$

$j = 1, \ldots, d$. Consequently,

$$(14.56) \qquad \sigma_j - \sigma_j^0 = \sum_{k=0}^d e_{jk}^t(x)\sigma_k^t,$$

where

$$(14.57) \qquad e_{jk}^t(0) = 0, \ j, k = 0, 1, \ldots, d,$$

and

$$(14.58) \qquad |e_{j0}^t(x)| \leq C(|x_0| + |x'|^2), \quad j = 1, \ldots, d.$$

Thus the functions $e_{\beta\gamma}^t$ satisfy the same estimates, (14.41), as the $e_{\beta\gamma}$, so they may be expressed in the form

$$(14.59) \qquad e_{\beta\gamma}^t(x) = \sum_{\langle\delta\rangle=|\beta|-\langle\beta\rangle+\langle\gamma\rangle} x^\delta e_{\beta\gamma\delta}^t(x), \ e_{\beta\gamma\delta}^t \in C^\infty.$$

Since

$$(14.60) \qquad x^{\alpha+\delta}e^{-i\langle x, \eta \rangle} = (-D_\eta)^{\alpha+\delta}e^{-i\langle x, \eta \rangle},$$

we may integrate by parts in (14.52) to obtain the integral

(14.61) $$\iint e^{-i\langle x,\eta\rangle} q_1^{(\alpha+\delta)}(0,\xi+\eta)[f_2]_\alpha^{(\beta;\gamma)}(tx,\sigma^t)e_{\beta\gamma\delta}^t(x)dx\,d\eta.$$

Now

(14.62) $$q_1^{(\alpha+\delta)} \in S_\mathcal{V}^{m_1-\langle\alpha\rangle-\langle\delta\rangle}(U),$$

and

(14.63) $$[f_2]_\alpha^{(\beta;\gamma)} \in \mathcal{F}^{m_2-\langle\beta\rangle+\langle\gamma\rangle}(U).$$

Setting $\sigma = \sigma^t$ in Proposition 14.45 we find that (14.61) is bounded by

$$C(t)(1+\|\xi\|)^{m-\langle\alpha\rangle-\langle\delta\rangle-\langle\beta\rangle+\langle\gamma\rangle}$$

(14.64)
$$\leq C(t)(1+\|\xi\|)^{m-|\alpha|-\langle\delta\rangle-\langle\beta\rangle+\langle\gamma\rangle}$$
$$= C(t)(1+\|\xi\|)^{m-N},$$

since $\langle\delta\rangle = |\beta| - \langle\beta\rangle + \langle\gamma\rangle$ and $|\alpha| + |\beta| = N$. $C(t)$ is bounded on $[0,1]$ and the $t$-integral is dominated by $(1+\|\xi\|)^{m-N}$, $m = m_1+m_2$. Corresponding estimates hold for derivatives of (14.61). Since $N$ is arbitrarily large the error produced by $r_N^{(2)}$ is, again, negligible.

(iii) Finally, the remainder term in (14.36) again has the form of (14.47) with

(14.65) $$q_1^{(\delta)} \in S_{\mathcal{V}^0}^{m_1-\delta}(U),$$

(14.66) $$r_{K,\delta}^{(3)}(x)[q_2]_\alpha^{(\beta;\gamma),0}(x,\xi) \in S_{\mathcal{V}^0}^{m_2-\langle\beta\rangle+\langle\gamma\rangle}(U),$$

where $\mathcal{V}^0$ denotes the hyperplane bundle generated by $X_1^0,\ldots,X_d^0$ in some neighborhood of $x = 0$. According to Proposition 14.45 the remainder in (14.36) is dominated by

(14.67) $$(1+\|\xi\|)^{m-\langle\delta\rangle-\langle\beta\rangle+\langle\gamma\rangle} \leq (1+\|\xi\|)^{m-K+\gamma_0},$$

where $K$ is arbitrarily large and $\gamma_0 \leq |\gamma| = |\beta| \leq N$. Corresponding estimates hold for the derivatives. This shows that the remainder in (14.36) is, again, negligible at $x = 0$.

We still need to prove that as a function of $y$, $q(0,\xi) \in \mathcal{F}^m(U)$. For the homogeneous parts of its asymptotic expansion, i.e. $q_r(0,\xi)$ as given by (14.8), this follows from Proposition 13.3. This leaves us with estimating $y$-derivatives of the neglected remainder terms. Let

(14.68) $$X_j(z) = \sum_{k=0}^{d} a_{jk}(z)\frac{\partial}{\partial z_k}, \quad j = 0,1,\ldots,d.$$

Then

(14.69) $$z = y + A(y)^t x$$

introduces the $y$-coordinates $x$, where $A(y) = (a_{jk}(y))$. Thus all the error terms depend on $y$ in the following manner:

(14.70) $$\iint e^{-i\langle x,\eta\rangle} f_1(y, \xi + \eta) f_2(y + A(y)^t x, \sigma(y; x, \xi)) dx \, d\eta,$$

where

(14.71) $$\sigma_j(y; x, \xi) = \xi_j + \sum_{k,l=0} c_{jkl}(y, x) x_k \xi_l,$$

$j = 0, 1, \ldots, d$, with $c_{jkl}(y, x)$ a smooth function of $y$ and $x$. Again, the proof of Proposition 13.3 shows that differentiating (14.70) with respect to $y$ leaves the original estimates, of the form (12.8), unchanged. This finishes the proof of Theorem 14.7.

(14.72) REMARKS: We could have used a result analogous to Theorem 2.9 of Hörmander [1] to show that as long as $q$ (i.e. $f$) is "tempered" and has an asymptotic expansion without conditions on the derivatives, it automatically satisfies "natural" estimates on the derivatives.

(14.73) MICROLOCALIZATION.: The calculus developed here is not "microlocal," since it does not contain the classical symbols of order zero, i.e. those with an expansion in terms which are homogeneous with respect to the standard (isotropic) dilations in the dual variables. It is easy to remedy this, since symbols of type $(\frac{1}{2}, \frac{1}{2})$ and symbols of type $(1, 0)$ compose according to the classical asymptotic formula: see Proposition 9.63. The algebra generated by $\mathcal{V}$-symbols and by classical symbols of order zero consists precisely of symbols having an asymptotic expansion of the type

$$q \sim \sum_{j,k=0}^{\infty} q'_{m-j} q''_{-k},$$

where $q'_{m-j}$ belongs to $S_{m-j,\mathcal{V}}(U)$ and $q''_{-k}$ is a classical symbol, homogeneous of order $-k$ in the dual variables. A somewhat more general mixed calculus is developed in Beals and Stanton [1].

§15 *Kernels of $\mathcal{V}$-Operators*

There is an asymptotic expansion of the distribution kernel of a $\mathcal{V}$-operator which corresponds to the asymptotic expansion of its symbol. To obtain its properties, we begin with a discussion of certain distributions on $\mathbf{R}^{d+1}$, characterized by their behavior under the Heisenberg dilation (11.37). The kernel of a pseudodifferential operator is formally given by $K(x, x - y)$, where

$$(15.1) \qquad\qquad K(x, z) = \int e^{i\langle z, \xi\rangle} q(x, \xi)\,\bar{d}\xi.$$

To make sense of the integral we think of $q$ as a tempered distribution on $\mathbf{R}^{d+1}_\xi$ whose singular support is at most $\{0\}$. As before, if $u \in C_c^\infty(\mathbf{R}^{d+1})$ and $\lambda > 0$, we define the dilation

$$(15.2) \qquad\qquad u_\lambda(\xi) = u(\lambda \cdot \xi), \quad \xi \in \mathbf{R}^{d+1},$$

see (11.37). This extends to distributions by setting

$$(15.3) \qquad\qquad \langle f_\lambda, u\rangle = \lambda^{-d-2}\langle f, u_{1/\lambda}\rangle, \quad f \in \mathcal{D}', \ u \in C_c^\infty.$$

If $f$ is a tempered distribution, then the relation between these dilations and the inverse Fourier transform, $\check{f}$, is

$$(15.4) \qquad\qquad (f_\lambda)^\vee = \lambda^{-d-2}(\check{f})_{1/\lambda}.$$

In particular, *if $f$ is homogeneous of degree $k$, i.e. $f_\lambda = \lambda^k f$, then $\check{f}$ is homogeneous of degree $-k - d - 2$.*

(15.5) DEFINITION: *For $k \in \mathbf{Z}$, $\mathcal{F}_k$ is the set of functions belonging to $C^\infty(\mathbf{R}^{d+1}\backslash 0)$ which are homogeneous of degree $k$ with respect to the dilations (15.2).*

(15.6) DEFINITION: *For $k \in \mathbf{Z}$, $\mathcal{G}_k$ denotes the set of distributions $g$, $g \in \mathcal{D}'(\mathbf{R}^{d+1})$, with the following properties: the singular support of $g$ is contained in $\{0\}$ and there are constants $c_\alpha$ such that*

$$(15.7) \qquad g_\lambda = \lambda^k g + \sum_{\langle\alpha\rangle = -k-d-2} c_\alpha(\lambda^k \log \lambda)\delta^{(\alpha)}, \quad \lambda > 0.$$

For purpose of normalization we fix a function $\phi \in C_c^\infty(\mathbf{R}^{d+1})$ with $\phi \equiv 1$ near the origin. As before, $\langle\alpha\rangle = \alpha_0 + |\alpha|$.

(15.8) PROPOSITION: *If $g \in \mathcal{G}_k$ then the restriction of $g$ to $\mathbf{R}^{d+1}\backslash 0$ belongs to $\mathcal{F}_k$. Conversely, if $f \in \mathcal{F}_k$ then there is a $g \in \mathcal{G}_k$ which agrees with $f$ on*

$\mathbf{R}^{d+1}\backslash 0$. *There is a unique such $g$ which satisfies*

$$(15.9) \qquad \langle g, \xi^\alpha \phi \rangle = 0 \quad if \quad \langle \alpha \rangle = -k - d - 2.$$

Proof. If $g \in \mathcal{G}_k$ then by assumption the restriction of $g$ to $\mathbf{R}^{d+1}\backslash 0$ is a smooth function $f$. Restricting (15.7) to $\mathbf{R}^{d+1}\backslash 0$, we see that $f$ is homogeneous of degree $k$.

Conversely, suppose $f \in \mathcal{F}_k$. Let $g$ be, for the moment, any distribution which agrees with $f$ on $\mathbf{R}^{d+1}\backslash 0$. For example the action of $g$ on $u \in C_c^\infty(\mathbf{R}^{d+1})$ may be defined by the absolutely convergent integral

$$(15.10) \quad \langle g, u \rangle = \int f(\xi) \left\{ u(\xi) - \sum_{0 \leq \langle \alpha \rangle \leq -k-d-2} (\alpha!)^{-1} u^{(\alpha)}(0) \xi^\alpha \phi(\xi) \right\} d\xi,$$

with $\phi$ as above. Then

$$(15.11) \qquad g_\lambda - \lambda^k g = \sum_{0 \leq \langle \alpha \rangle \leq -d-2-k} c_\alpha(\lambda) \delta^{(\alpha)},$$

where

$$c_\alpha(\lambda) = \frac{\lambda^k}{\alpha!} \int f(\xi) \xi^\alpha [\phi(\xi) - \phi(\lambda \cdot \xi)] d\xi.$$

Assuming

$$\phi(\xi) = \psi(\|\xi\|) = \psi\left( \left[ \xi_0^2 + \left( \sum_{j=1}^d \xi_j^2 \right)^2 \right]^{\frac{1}{4}} \right),$$

we obtain

$$c_\alpha(\lambda) = \frac{\lambda^k}{\alpha!} \int_{\|\theta\|=1} f(\theta) \theta^\alpha d\theta \int_0^\infty (\psi(r) - \psi(\lambda r)) r^{d+1+k+\langle \alpha \rangle} dr.$$

This easily yields

$$(15.12) \qquad c_\alpha(\lambda) = \begin{cases} c_\alpha(\lambda^k - \lambda^{-d-2-\langle \alpha \rangle}), & \text{if} \quad \langle \alpha \rangle < -d - 2 - k, \\[2mm] c_\alpha \lambda^k \log \lambda, & \text{if} \quad \langle \alpha \rangle = -d - 2 - k, \end{cases}$$

where

$$(15.13) \qquad c_\alpha = \begin{cases} \dfrac{1}{\alpha!(d+2+k+\langle \alpha \rangle)} \displaystyle\int_0^\infty x^{d+2+k+\langle \alpha \rangle} \; \psi'(x) dx \displaystyle\int_{\|\theta\|=1} f(\theta) \theta^\alpha d\theta \\[2mm] \qquad\qquad\qquad\qquad\qquad\qquad \text{if} \quad \langle \alpha \rangle < -d - 2 - k, \\[2mm] \dfrac{1}{\alpha!} \displaystyle\int_0^\infty \phi'(x) dx \displaystyle\int_{\|\theta\|=1} f(\theta) \theta^\alpha d\theta \quad \text{if} \quad \langle \alpha \rangle = -d - 2 - k. \end{cases}$$

Thus (15.11) has the form

$$g_\lambda = \lambda^k g + \sum_{0 \leq \langle \alpha \rangle < -d-2-k} c_\alpha [\lambda^k - \lambda^{-d-2-\langle \alpha \rangle}] \delta^{(\alpha)}$$

(15.14)

$$+ \sum_{\langle \alpha \rangle = -d-2-k} c_\alpha (\lambda^k \log \lambda) \delta^{(\alpha)}.$$

To kill the first sum on the right-hand side of (15.14), we replace $g$ by

(15.15)
$$g + \sum_{\langle \alpha \rangle < -d-2-k} c_\alpha \delta^{(\alpha)}.$$

This produces an element of $\mathcal{G}_k$, as desired. Moreover this element is unique up to a distribution with support $\{0\}$, which must have the form

(15.16)
$$\sum_{\langle \alpha \rangle = -d-2-k} a_\alpha \delta^{(\alpha)}, \quad a_\alpha \text{ constant.}$$

The $a_\alpha$ are uniquely determined by the supplementary condition (15.9).

An interesting consequence of (15.13)–(15.15) is that $f \in \mathcal{F}_k$ can be *extended to a homogeneous distribution on* $\mathbf{R}^{d+1}$ *if and only if*

(15.17)
$$\int_{\|\theta\|=1} f(\theta) \theta^\alpha d\theta = 0, \quad \langle \alpha \rangle = -k - d - 2.$$

(15.18) DEFINITION: *We shall denote by* $\overset{\circ}{\mathcal{F}}_k$ *that subspace of* $\mathcal{F}_k$ *whose elements satisfy (15.17).*

Note that if $g$ belongs to $\mathcal{G}_k$, then it has polynomial growth at $\infty$ and is therefore a tempered distribution. We now examine its inverse Fourier transform.

(15.19) DEFINITION: *For* $j \in \mathbf{Z}$, $\mathcal{K}_j$ *denotes the set of distributions* $K \in \mathcal{D}'(\mathbf{R}^{d+1})$ *with the properties: the singular support of* $K$ *is contained in* $\{0\}$ *and there are constant* $c_\alpha$ *such that*

(15.20)
$$K_\lambda = \lambda^j K + \sum_{\langle \alpha \rangle = j} c_\alpha (\lambda^j \log \lambda) x^\alpha.$$

(15.21) PROPOSITION: *Let* $K \in \mathcal{K}_j$. *Then*

(15.22)
$$K(x) = f(x) + p(x) \log \|x\|, \quad x \neq 0,$$

*where* $f \in \mathcal{F}_j$, *p is a polynomial, homogeneous of degree j with respect to the dilations (11.37), and*

(15.23)
$$\|x\| = \left( x_0^2 + \left( \sum_{j=1}^{d} x_j^2 \right)^2 \right)^{\frac{1}{4}}.$$

Proof. Equation (15.20) makes sense pointwise for $x \neq 0$. Therefore we may replace $x$ by $x\|x\|^{-1}$ and $\lambda$ by $\|x\|$ in it. This gives (15.22) with $f$ as the homogeneous function which agrees with $K$ on $\{\|x\| = 1\}$ and $p$ as the polynomial with coefficients $c_\alpha$, $\langle \alpha \rangle = j$.

(15.24) PROPOSITION: *The inverse Fourier transform is a bijection from $G_k$ to $\mathcal{K}_{-k-d-2}$.*

Proof. It follows from (15.4) that the inverse Fourier transform is a bijection from tempered distributions satisfying (15.7) to tempered distributions satisfying (15.20) with $j = -k - d - 2$. It remains to check the singular supports. Let $g \in \mathcal{G}_k$ and write $g = \phi f + (1 - \phi)g$. $(\phi g)^\vee$ is analytic. Next, for every multi-index $\alpha$, $D^\beta(\xi^\alpha[1 - \phi]g)$ is integrable at $\infty$ for every $\beta$ with $\langle \beta \rangle$ sufficiently large. It follows that $x^\beta D^\alpha([(1 - \phi)g]^\vee)$ is continuous. Thus $\check{g}$ has singular support at most $\{0\}$. Conversely, the same argument shows that for $K \in \mathcal{K}_j$, $\hat{K}$ has singular support at most $\{0\}$.

Next we topologize $\mathcal{G}_k$ and $\mathcal{K}_j$:
(15.25) If $k \in \mathbf{Z}$, $k > -d - 2$, $g$ is uniquely determined by its restriction to $\mathbf{R}^{d+1}\backslash 0$. Thus we may assign $\mathcal{G}_k$ the topology induced from $C^\infty(\mathbf{R}^{d+1}\backslash 0)$, or equivalently the topology of $C^\infty(\Omega_1)$, where $\Omega_1$ is the unit Heisenberg sphere.
(15.26) If $k \in \mathbf{Z}$, $k \leqq -d - 2$, the map

$$g \rightarrow (g|_{\mathbf{R}^{d+1}\backslash 0}, \{\langle g, \xi^\alpha \phi \rangle : \langle \alpha \rangle = -k - d - 2\})$$

is a bijection from $\mathcal{G}_k$ to $\mathcal{F}_k \times \mathbf{C}^{N_k}$, where $N_k$ denotes the cardinality of the index set $\{\alpha \in \mathbf{Z}_+^{d+1} : \langle \alpha \rangle = -k - d - 2\}$. Consequently we may topologize $\mathcal{G}_k$ by the topology induced by $C^\infty(\mathbf{R}^{d+1}\backslash 0) \times \mathbf{C}^{N_k}$, or equivalently, by $C^\infty(\Omega_1) \times \mathbf{C}^{N_k}$.
As for $\mathcal{K}_j$:
(15.27) $j \geqq 0$. The map

$$K \rightarrow (K_{\mathbf{R}^{d+1}\backslash 0}, \{c_\alpha\}),$$

where the $c_\alpha$'s are the constants given by (15.20), is a bijection of $\mathcal{K}_j$ onto $\mathcal{F}_j \times \mathbf{C}^{N_k}$, $j + k = -d - 2$, and $\mathcal{K}_j$ may be topologized by $C^\infty(\Omega_1) \times \mathbf{C}^{N_k}$.
(15.28) $-d - 2 < j < 0 : \mathcal{K}_j$ may be topologized by $C^\infty(\Omega_1)$.

(15.29) $j \leq -d - 2$: as in (15.26) the map

$$K \rightarrow (K_{\mathbf{R}^{d+1}\backslash 0}, \{\langle k, \xi^\alpha \hat\phi \rangle : \langle \alpha \rangle = -j - d - 2\})$$

is a bijection from $\mathcal{K}_j$ to $\overset{\circ}{\mathcal{F}}_j \times \mathbf{C}^{N_k}$, thus $\mathcal{K}_j$ may be topologized by identi-
fication with $\{\overset{\circ}{\mathcal{F}}|_{\|\xi\|=1}\} \times \mathbf{C}^{N_j} \subset C^\infty(\Omega_1) \times \mathbf{C}^{N_j}$.

Now the proof of Proposition 15.24 also yields

(15.30) PROPOSITION: *The inverse Fourier transform is a homeomor-
phism from $\mathcal{G}_k$ onto $\mathcal{K}_{-k-d-2}$, where the topologies for $\mathcal{G}_k$ and $\mathcal{K}$ are given
by (15.25)–(15.29).*

Suppose now that $U \subset M$ carries a hyperplane bundle $\mathcal{V}$ and corresponding vector fields $\{X_j\}$. Given $y \in U$, denote the $y$-coordinate map
by

(15.31)                              $\psi_y : U \rightarrow \mathbf{R}^{d+1}$.

If $y \rightarrow K_y$ is a smooth map from $U$ to the Frechet space $\mathcal{K}_j$, then one may
define the distribution $K \in \mathcal{D}'(U \times U)$ by setting

(15.32)              $\langle K, v \rangle = \int \langle K_y, v_y' \rangle dy, v \in C_c^\infty(U \times U),$

where

(15.33)              $v_y(x) = v(y, \psi_y^{-1}(x)), x \in \mathbf{R}^{d+1},$

and

(15.34)                              $v'(x) = v(-x).$

(15.35) DEFINITION: *For $j \in \mathbf{Z}$, $\mathcal{K}_j(U)$ denotes the set of distributions
$K \in \mathcal{D}'(U \times U)$, which are given by (15.32)–(15.34), with $y \rightarrow K_y$ a smooth
map from $U \rightarrow \mathcal{K}_j$.*

(15.36) PROPOSITION: *Suppose $K \in \mathcal{K}_j(U)$. Then the singular support of
$K$ is the diagonal $\Delta = \{(y, y)\}$. On the complement of $\Delta$, $K$ has the form:*

(15.37)          $K(y, x) = f(y, -\psi_y(z)) + p(y, -\psi_y(z)) \log \|\psi_y(z)\|,$

*where $f(y, \cdot)$ is $H$-homogeneous of degree $j$ and $(p(y, \cdot)$ is an $H$-homogene-
ous polynomial, also of degree $j$.*

Proof. The assumption on the mapping $y \rightarrow K_y$ implies that the function
$F(y, x) = K_y(x)$ is smooth on $U \times (\mathbf{R}^{d+1}\backslash 0)$. Since $\psi_y$ is an affine map, its

Jacobian is a non-vanishing function of $y$ alone. It follows readily that on the complement of $\Delta$, $K$ is given by a kernel

(15.38) $$K(y, z) = |\psi'_y|^{-1} F(y, -\psi_y(z)).$$

In view of Proposition 15.21 this is of the form (15.37).

We come now to the principal result of this section.

(15.39) THEOREM: *Suppose $Q$ is a $\mathcal{V}$-operator of degree $m$, $m \in \mathbf{Z}$. Then the Schwartz kernel $K$ of $Q$ has an asymptotic expansion,*

(15.40) $$K \sim \sum_{j \geq -d-2-m} K_j, \quad K_j \in \mathcal{K}_j(U),$$

*in the sense that for any $N \geq 0$ there is an integer $J$ such that the distribution*

(15.41) $$K - \sum_{j \leq J} K_j$$

*belongs to $C^N(U \times U)$.*

Proof. Let $\sum_{j \geq 0} q_{m-j}$ denote the asymptotic expansion of the symbol of $Q$, so

$$q_k(y, \xi) = f_k(y, \sigma(y, \xi)), \quad f_k \in \mathcal{F}_k(U).$$

Choose a function $\eta \in C^\infty(\mathbf{R}^{d+1})$, with $\eta \equiv 0$ near the origin and $\eta \equiv 1$ near $\infty$, and set

(15.42) $$\tilde{q}_k(y, \xi) = q_k(y, \xi)\eta(\sigma(y, \xi)), \quad Q_k = \tilde{q}_k(y, D).$$

Given $y \in U$, in the $y$-coordinates one has

(15.43) $$\begin{aligned} Q_k u(0) &= \int f_k(y, \xi)\eta(\xi)\hat{u}(\xi)\,\bar{d}\xi \\ &= \langle [f_k(y, \cdot)\eta]^\vee, u' \rangle, \end{aligned}$$

since in the $y$-coordinates $\sigma(0, \xi) = \xi$. Now $f_k(y, \cdot) \in \mathcal{F}_k$ and can be extended, as in Proposition 15.8, to a distribution $g_{k,y} \in \mathcal{G}_k$. This differs from the distribution $f_k(y, \cdot)\eta(\cdot)$ by a compactly supported distribution, so

(15.44) $$Q_k u(0) = \langle K_{j,y}, u' \rangle + \langle L_{j,y}, u' \rangle,$$

where $j + k = -d - 2$, $K_{j,y} \in \mathcal{K}_{-d-2-k}$ and $L_{j,y} \in C^\infty(\mathbf{R}^{d+1})$. Thus in the original coordinates

(15.45) $$Q_k v(y) = \langle K_{j,y}, v' \circ \psi_y^{-1} \rangle + \langle L_{j,y}, v' \circ \psi_y^{-1} \rangle, \quad v = u \circ \psi_y.$$

The dependence on $y$ is smooth, so the Schwartz kernel of the operator $Q_k$ has the form $K_{-d-2-k} + L_{-d-2-k}$ with $K_j \in \mathcal{K}_j(U)$ and $L_j \in C^\infty(U \times U)$. The operator

(15.46)
$$Q - \sum_{j<N} Q_{m-j}$$

has degree $m - N$, so its kernel becomes progressively smoother as $N \to \infty$. It follows that the kernel of $Q$ has the expansion (15.40) with $K_j$ obtained from the term $Q_k$, $j + k = -d - 2$.

(15.47) REMARK: Set

(15.48)
$$d(y, z) = \|\psi_y(z)\|,$$

where again $\psi_y$ is the $y$-coordinate map and $\|\cdot\|$ is given by (15.23). This is a quasi-metric on $U$ and, together with the Haar measure $dx$, gives $U$ the structure of a space of homogeneous type (Coifman and Weiss [1]). This, together with the information on the kernels obtained above, gives the necessary real variable information for the study of the mapping properties of $\mathcal{V}$-operators with respect to $L^p$ and Hölder spaces. For example, a $\mathcal{V}$-operator of degree 0 maps $L^p_c(U)$ to $L^p_{\text{loc}}(U)$, $1 < p < \infty$. The $L^p$ and Hölder regularity theory of the second order hypoelliptic operators of §18 and of $\Box_b$ of Chapter 4 is a consequence; cf. Beals [1], Nagel and Stein [2], Beals, Greiner, and Stanton [3].

The proof of Theorem 15.39 is reversible and gives the converse result.

(15.49) THEOREM: *Suppose $Q : C^\infty_c(U) \to C^\infty(U)$ has Schwartz kernel $K$ which has an asymptotic expansion (15.40) with $K_j \in \mathcal{K}_j(U)$. Then $Q$ is a $\mathcal{V}$-operator of degree $m$.*

Proof. Let $j = -d - 2 - k$ and define

(15.50)
$$Q_k v(y) = \langle K_{j,y}, v' \circ \psi_y^{-1} \rangle, \quad v \in C^\infty_c(U).$$

Then by assumption (15.40) becomes progressively more smoothing as $N \to \infty$, so it suffices to examine $Q_k$. As in (15.43), in the $y$-coordinates

(15.51)
$$Q_k u(0) = \langle g_{k,y}, \hat{u} \rangle,$$

where $g_{k,y} \in \mathcal{G}_k$ is the Fourier transform of $K_{j,y}$. Set

(15.52)
$$q_k(y, \xi) = f_k(y, \sigma(y, \xi)) = g_{k,y}(\sigma(y, \xi)), \quad \xi \neq 0.$$

Then $f_k \in \mathcal{F}_k(U)$, so $q_k \in S_{k,\mathcal{V}}(U)$. The operator $Q_k$ is defined by the symbol $q_k$ suitably regularized at $\xi = 0$, and it follows that $Q_k$ is a $\mathcal{V}$-operator with the single term $q_k$ in the asymptotic expansion of its symbol.

Since the operator $Q_k$ has kernel $K_j$, $j + k = -d - 2$, modulo a smoothing operator, we have derived Theorem 15.49.

§16 *The Proof of the Invariance Theorem (Theorem 10.67)*

Let $U$ and $\tilde{U}$ denote open subsets of $\mathbf{R}^{d+1}$, carrying hyperplane bundles $\mathcal{V}$ and $\tilde{\mathcal{V}}$, and suppose $\phi : U \to \tilde{U}$ is a diffeomorphism carrying $\mathcal{V}$ to $\tilde{\mathcal{V}}$. Let $\tilde{Q}$ denote a $\tilde{\mathcal{V}}$-operator on $\tilde{U}$ and $Q$ its pullback to $U$ by $\phi$, and let $\tilde{K}$ denote the kernel of $\tilde{Q}$ and $K$ its pullback to $U$, so $K$ is the kernel of $Q$. Then $K$ is obtained from (15.32) with $K_y$ defined on $\psi_y(U)$ by

(16.1)
$$\begin{aligned}
\langle K_y, u \rangle &= Q(u \circ \psi_y)(y) \\
&= \tilde{Q}'(u \circ \psi_y \circ \phi^{-1})(\phi(y)) \\
&= \langle \tilde{K}_{\tilde{y}}, u \circ \psi_y \circ \phi^{-1} \circ (\tilde{\psi}_{\tilde{y}})^{-1} \rangle \\
&= \langle \tilde{K}_{\tilde{y}}, u \circ \phi_y^{-1} \rangle.
\end{aligned}$$

Here $\tilde{y} = \phi(y)$, $\tilde{\psi}_{\tilde{y}}$ denotes the corresponding coordinate map and $\phi_y$ is defined by

(16.2)
$$\begin{array}{ccc}
U & \xrightarrow{\;\;\phi\;\;} & \tilde{U}' \\
{\scriptstyle\psi_y}\downarrow & & \downarrow{\scriptstyle\tilde{\psi}_{\tilde{y}}} \\
\mathbf{R}^{d+1} & \xrightarrow[\;\;\phi_y\;\;]{} & \mathbf{R}^{d+1}
\end{array}$$

We use the following systems of coordinates:

(16.3)
$$x = \psi_y(z), \quad \tilde{x} = \tilde{\psi}_{\tilde{y}}, \quad \tilde{z} = \psi(z).$$

In the $y$-coordinates

(16.4)
$$X_j = \frac{\partial}{\partial x_j} + \sum_{k=0}^{d} O(|x|)\frac{\partial}{\partial x_k}, \quad j = 0, 1, \ldots, d.$$

To say that $\phi$ takes $\mathcal{V}$ to $\tilde{\mathcal{V}}'$ implies that in the $\tilde{y}$-coordinates

(16.5)
$$X_j = \frac{\partial}{\partial \tilde{x}_j} + \sum_{k=0}^{d} O(|\tilde{x}|)\frac{\partial}{\partial \tilde{x}_k}, \quad j = 1, \ldots, d.$$

This is equivalent to

(16.6)
$$\tilde{x}_0 = x_0 + O(|x|^2),$$

(16.7) $\qquad \tilde{x}_j = x_j + O(|x_0|) + O(|x|^2), \quad j = 1, \ldots, d.$

We say that *a diffeomorphism $\phi$ between neighborhoods of the origin of* $\mathbf{R}^{d+1}$ *is an admissible diffeomorphism if it satisfies (16.6) and (16.7).* In the ensuing discussion we suppress the parameter $y \in U$ and study the behavior of sums of $H$-homogeneous distributions under the action of an admissible diffeomorphism.

Let $\mathcal{K}_j$ be the space of distributions of Definition 15.19. Note that for $j > -d - 2$ an element of $\mathcal{K}_j$ is defined by its restriction to $\mathbf{R}^{d+1}\backslash(0)$. Furthermore

(16.8) $\qquad \mathcal{K}_j \subset C^{j-1}(\mathbf{R}^{d+1}), \quad j \geq 1.$

Note also the effects of differentiation and of multiplication by a monomial $x^\alpha$:

(16.9) $\qquad D^\alpha : \mathcal{K}_j \to \mathcal{K}_{j-\langle\alpha\rangle},$

(16.10) $\qquad x^\alpha : \mathcal{K}_j \to \mathcal{K}_{j+\langle\alpha\rangle}.$

(16.11) DEFINITION: *For $m \in \mathbf{Z}$, $\mathcal{K}^m$ is the set of distributions $K \in \mathcal{D}'(\mathbf{R}^{d+1})$ having an asymptotic expansion of the form*

(16.12) $\qquad K \sim \sum_{k \geq 0} K_{m+k}, \quad K_j \in \mathcal{K}_j,$

*in the sense that for every integer $N > 0$ there is an $M > 0$, such that*

(16.13) $\qquad K - \sum_{k < M} K_{m+k} \in C^N(\mathbf{R}^{d+1}).$

Note that a consequence of (16.13) is that the singular support of $K$, like that of $K_j$, is empty or $\{0\}$.

(16.14) PROPOSITION: *$\mathcal{K}^m$ is a module over $C^\infty(\mathbf{R}^{d+1})$.*

Proof. Suppose $K \in \mathcal{K}^m$ and $f \in C^\infty(\mathbf{R}^{d+1})$. Set

(16.15) $\qquad L_k = \sum_{\langle\alpha\rangle+j=k} \frac{1}{\alpha!} f^{(\alpha)}(0) x^\alpha K_j = \sum f_\alpha K_j.$

By (16.10) $L_k$ belongs to $\mathcal{K}_k$. Now

(16.16)
$$fK - \sum_{k < 2M} L_{m+k} = \left( f - \sum_{\langle\alpha\rangle < M} f_\alpha \right) K$$
$$+ \left( \sum_{\langle\alpha\rangle < M} f_\alpha \right) \left( K - \sum_{k < M} K_{m+k} \right) + R_M,$$

where $R_m \in K^{m+M}$. As $M \to \infty$, the first term on the right becomes smooth because of Taylor's Theorem, the second because of (16.13) and $R_M$ because of (16.8).

This result allows us to localize in the usual way. *If $\Omega$ is a neighborhood of 0 in $\mathbf{R}^{d+1}$, we set*

$$(16.17) \qquad \mathcal{K}_\Omega^m = \{K \in \mathcal{D}'(\Omega); \chi K \in \mathcal{K}^m \text{ if } \chi \in C_c^\infty(\Omega)\}$$

*Let $\phi$ denote a diffeomorphism, $\phi : U \to U$. Given a distribution $f \in \mathcal{D}'(U)$, we define $f_\phi \in \mathcal{D}'(U)$ by*

$$(16.18) \qquad \langle f_\phi, u \rangle = \langle f, u \circ \phi \rangle, \quad u \in C_c^\infty(U).$$

(16.19) THEOREM: *Suppose $\phi$ is an admissible diffeomorphism, $\phi : U \to U$, $U$ a neighborhood of the origin in $\mathbf{R}^{d+1}$. Let $K \in \mathcal{K}^m$. Then $K_\phi \in \mathcal{K}_\Omega^m$, for some neighborhood $\Omega$ of the origin in $\mathbf{R}^{d+1}$, $\Omega \subset U$.*

Proof. Suppose $\phi$ has the form

$$(16.20) \qquad \phi(x) = \phi(x_0, x') = (x_0 + \varepsilon(x'), x'),$$

where $\varepsilon$ is quadratic. Then $\phi$ commutes with $H$-dilation and therefore preserves both $\mathcal{K}_j$ and $\mathcal{K}^m$. Composing with a diffeomorphism of type (16.20), the general admissible diffeomorphism may be reduced to one of the form

$$(16.21) \qquad \phi(x_0, x') = (x_0 + \varepsilon_0(x), x' + \varepsilon'(x)),$$

where

$$(16.22) \qquad |\varepsilon_0(x) \leq C(x_0^2 + |x_0| \, |x'| + |x'|^3),$$

$$(16.23) \qquad |\varepsilon'(x)| \leq C(|x_0| + |x'|^2).$$

Given $u \in C_c^\infty(U)$, consider the Taylor expansion

$$(16.24) \qquad (u \circ \phi)(x) \sim \sum_\alpha (\alpha!)^{-1} u^{(\alpha)}(x) \varepsilon(x)^\alpha,$$

where we set $\varepsilon(x) = (\varepsilon_0(x), \varepsilon'(x))$. Then

$$(16.25) \qquad \begin{aligned} \langle K_\phi, u \rangle &= \sum_{|\alpha| < N} (\alpha!)^{-1} \langle K, \varepsilon^\alpha u^{(\alpha)} \rangle + \langle r_N, u \rangle \\ &= \sum_{|\alpha| < N} (\alpha!)^{-1} (-1)^\alpha \langle \partial^\alpha [\varepsilon^\alpha K], u \rangle + \langle r_N, u \rangle. \end{aligned}$$

In view of (16.22) and (16.23), $\varepsilon(x)^\alpha$ can be written in the form

(16.26) $$\varepsilon(x)^\alpha = \sum_\beta h_{\alpha\beta}(x)x^\beta,$$

with $h_{\alpha\beta}(x) \in C^\infty(U)$ and $\langle\beta\rangle \geq \langle\alpha\rangle + |\alpha| \geq \frac{3}{2}\langle\alpha\rangle$. From this, (16.9), (16.10) and Proposition 16.14 we see that $\partial^\alpha[\varepsilon^\alpha K] \in \mathcal{K}^{m(\alpha)}$, with $m(\alpha) \leq m + \frac{1}{2}\langle\alpha\rangle$. It remains to show that $r_N$ is progressively smoother as $N \to \infty$. Now $\langle r_N, u\rangle$ is a linear combination of integrals over $s \in [0,1]$ of terms

(16.27)     $\langle K, u^{(\alpha)}(x + s\varepsilon(x))\varepsilon(x)^\alpha\rangle = \langle \varepsilon^\alpha K, u^{(\alpha)} \circ \phi_s\rangle, \quad |\alpha| = N.$

Localizing again, if necessary, we may assume that $\phi_s$ is a diffeomorphism for $0 \leq s \leq 1$. For $|\alpha|$ large, the distribution $\varepsilon^\alpha K$ is function which is smooth of order $m + |\alpha| + \langle\alpha\rangle$. Changing variables and integrating by parts, we find that $r_N$ is smooth of order $m + N$. This completes the proof of Theorem 16.19.

(16.28) *End of proof of Theorem 10.67.* Let $\tilde{Q}$ be a $\tilde{\mathcal{V}}$-operator with kernel $\tilde{K}' \in \mathcal{K}^m(\tilde{U})$. By Theorem 15.39 $\tilde{K}_{\tilde{y}} \in \mathcal{K}^m$ in $\tilde{y}$-coordinates. Suppose $\phi$ is a diffeomorphism, $\phi : U \to \tilde{U}$. Then $K_y$, the pullback of $\tilde{K}_{\tilde{y}}$, $\tilde{y} = \phi(y)$, is given by (16.1) in $y$-coordinates. Theorem 16.19 shows that $K_y \in \mathcal{K}^m_\Omega$, for some neighborhood $\Omega$ of the origin. $K_y$ depends smoothly on $y$, as can be seen by the proof of Theorem 16.19, so Theorem 15.49 implies that $Q$, with kernel $K$, is a $\mathcal{V}$-operator, $\mathcal{V} = \phi^*\tilde{\mathcal{V}}$.

## §17 *Adjoints of $\mathcal{V}$-Operators*

Let $\mathcal{K}^m(U)$, $m \in \mathbf{Z}$, denote the space of distributions $K \in \mathcal{D}'(U \times U)$, such that $K \in \mathcal{K}^m(U)$ if and only if

(17.1) $$K \sim \sum_{j \geq m} K_j, \quad K_j \in \mathcal{K}_j(U),$$

in the sense that, for any $N \in \mathbf{Z}_+$, there is $J \in \mathbf{Z}_+$, so that

(17.2) $$K - \sum_{m \leq j \leq J} K_j \in C^N(U \times U).$$

According to Theorem 15.39 $Q$ is a $\mathcal{V}$-operator of degree $m$, $m \in \mathbf{Z}$, if and only if the distribution kernel of $Q$ belongs to $\mathcal{K}^{-m-d-2}(U)$.

Let $\rho(x)dx$ denote a smooth positive density on $U \subset M$. Relative to this density the formal adjoint $Q^*$ of $Q$ is given by

$$(17.3) \qquad \langle Q^*u, v \rangle_\rho = \int u(y)\overline{Qv(y)}\rho(y)dy, \quad u, v \in C_c^\infty(U).$$

(17.4) THEOREM: *If $Q$ is a $\mathcal{V}$-operator of degree $m$, $m \in \mathbb{Z}$, then so is its formal adjoint $Q^*$. The principal symbol of $Q^*$ is the complex conjugate of the principal symbol of $Q$.*

Proof. With the usual formal notation, if $Q$ has distribution kernel $K$, then $Q^*$ has distribution kernel $K^*$, where

$$(17.5) \qquad K^*(y, z) = \overline{K(z, y)}\, \frac{\rho(z)}{\rho(y)}.$$

It suffices to consider the case of a single term in the asymptotic expansion (17.1). In that case

$$(17.6) \qquad K^*(y, z) = \frac{1}{\rho(y)} \overline{F}(z, -\psi_z(y)),$$

where $z \to F(z, \cdot)$ is a smooth map from $U$ to some $\mathcal{K}_j$. Taking the Taylor expansion with respect to $z$ we have

$$(17.7) \qquad K^*(y, z) = \frac{1}{\rho(y)} \sum_{|\alpha| < N} \frac{1}{\alpha!} \overline{F}^{(\alpha)}(y, -\psi_z(y))(z - y)^\alpha + r_N(y, z),$$

where $r_N$ is smooth for $y \neq z$ and is $O(|y - z|^{j/2+N})$ for large $N$, with similar estimates for its derivatives (if $j < 0$, replace $j/2$ by $j$). Thus

$$(17.8) \qquad K^*(y, z) \sim \frac{1}{\rho(y)} \sum_\alpha \overline{F}^\alpha(y, -\psi_z(y))(z - y)^\alpha / \alpha!.$$

Now $\rho(y)^{-1}$ is harmless and we only need to consider a single term of (17.8):

$$(17.9) \qquad \overline{F}^{(\alpha)}(y, -\psi_z(y))(z - y)^\alpha,$$

where $\overline{F}^{(\alpha)}(y, \cdot)$ is a smooth map from $U$ to $\mathcal{K}_j$. We write

$$(17.10) \qquad \psi_z(y) = A_z(y - z) = -A_z A_y^{-1} \psi_y(z),$$

where $A_z$ and $A_y$ are linear. Thus we may write

$$(17.11) \qquad \overline{F}^{(\alpha)}(y, -\psi_z(y)) = \overline{F}^{(\alpha)}(y, \phi_y(\psi_y(z)))$$

where

$$(17.12) \qquad \phi_y(w) = A_{\psi_y^{-1}(w)} A_y^{-1}(w).$$

Localizing (if necessary), we may assume that $\phi_y$ is a diffeomorphism. Note that $\phi_y(0) = 0$ and $\phi_y'(0) = I$. Thus $\phi_y$ is a smooth family of admissible diffeomorphisms, and by the proof of Theorem 16.19

$$(17.13) \qquad y \to F^{(\alpha)}(y, \phi_y(\cdot))$$

is a smooth map: $U \to \mathcal{K}^j(U)$. To finish the argument we note that

$$(17.14) \qquad \psi_y(z) = A_y(z - y),$$

hence

$$(17.15) \qquad z - y = A_y^{-1}(\psi_y(z)).$$

Consequently

$$(17.16) \qquad y \to F^{(\alpha)}(y, \phi_y(\cdot))[A_y^{-1}(\cdot)]^\alpha$$

is still a smooth map to $\mathcal{K}^{j+(\alpha)}(U)$. Thus $K^* \in \mathcal{K}^j(U)$ if $K \in \mathcal{K}_j(U)$. Therefore

$$(17.17) \qquad K \in \mathcal{K}^{-m-d-2}(U) \Longrightarrow K^* \in \mathcal{K}^{-m-d-2}(U).$$

Finally, *the "principal part" of $K^*$ is given by*

$$(17.18) \qquad \overline{K_{-m-d-2}(z, -x)}$$

*if $K_{-m-d-2}(z, x)$ denotes the "principal part" of $K$.* Then the Fourier transform of (17.18) in the second variable yields the last statement of Theorem 17.4.

§18 *Hypoellipticity and Parametrices for Second Order Differential Operators*

In this section we return to the study of the second order operators of §1. We use the results of Chapters 1 and 2 and of this chapter to obtain our goal: a sharp version of Theorem 2.3 relating hypoellipticity to invertibility of the related model operators.

Suppose $U$ is a domain in $\mathbf{R}^{d+1}$ and $P$ is an operator of the form (1.7), i.e.

$$(18.1) \qquad P = -\sum_1^d X_j^2 - i\lambda X_0 + \left\{ \sum_1^d \gamma_j X_j + c \right\},$$

$\lambda, \gamma_j, c \in C^\infty(U)$, $X_0, X_1, \ldots, X_d$ independent vector fields on $U$. Let $\mathcal{V} \subset TU$ be the subbundle generated by $X_1, \ldots, X_d$. Given $y \in U$, the model operator $P^y$ of Definition 1.12,

$$(18.2) \qquad P^y = -\sum_1^d (X_j^y)^2 - i\lambda(y)X_0^y,$$

is precisely the $y$-invariant operator associated with the principal symbol of $P$ in $S_\mathcal{V}^2(U)$, since the term in braces in (18.1) has symbol belonging to $S_\mathcal{V}^1(U)$.

(18.3) DEFINITION: *A parametrix for $P$ is an operator $Q : C_c^\infty(U) \to C^\infty(U)$ with the property that the operators $QP - I$ and $PQ - I$ are smoothing.*

We may now formulate the main result of §18:

(18.4) THEOREM: *The following are equivalent conditions for the operator $P$ of (18.1):*

   (a) *For each $y \in U$, $\lambda(y)$ does not belong to the singular set $\Lambda_y$ of Definition 2.4.*

   (b) *Each model operator $P^y$ has a pseudodifferential inverse $Q^y$.*

   (c) *$P$ has a pseudodifferential parametrix $Q$ with symbol $q \in S_\mathcal{V}^{-2}(U)$.*

   (d) *$P$ is hypoelliptic with loss of one derivative, i.e. if $u \in \mathcal{D}'(U)$ and $Pu \in H_{\mathrm{loc}}^s(U)$, then necessarily $u \in H_{\mathrm{loc}}^{s+1}(U)$.*

*Moreover, when these conditions hold, the principal symbol $q_{-2}$ of the parametrix is related to the symbol $q^y$ of the model opeator $Q^y$ by (11.47) and (11.48), where*

$$(18.5) \qquad q_{-2}(y; 0, \xi) = q^y(0, \xi), \quad y \in U, \quad \xi \in \mathbf{R}^{d+1}.$$

Proof. We showed in Chapter 2 that (a) $\Longrightarrow$ (b). The plan of the remainder of the proof is to show (b) $\Longrightarrow$ (c) $\Longrightarrow$ (d) $\Longrightarrow$ (a).

   (i) *Proof that (b) $\Longrightarrow$ (c).* We showed in Chapter 2 that in the $y$-coordinates a pseudodifferential inverse $Q^y$ must have a symbol of the form

$$(18.6) \qquad q^y(x, \xi) = g(y; \sigma^y(x, \xi))$$

where $g(y, \xi) = q_\lambda(\xi)$ defined in Theorem 8.60, is $H$-homogeneous of degree $-2$. Moreover, $g(y, \xi)$ depends smoothly on $y$ since it depends smoothly on the associated matrix $B = (b_{jk})$ of (1.11); see the discussion in Remark 8.91. Therefore (18.5) defines a symbol $q_{-2} \in S_{-2,\mathcal{V}}(U)$. Let $Q_{-2}$ denote an associated operator whose symbol belongs to $S_\mathcal{V}^{-2}(U)$ and has $q_{-2}$ as the only term in its asymptotic expansion. We may assume that $Q_{-2}$ is properly supported (see Remark 10.79). Let $p_2$ be the principal symbol of $P$, so that $p_2^y$ is the symbol of the model operator $P^y$. By construction, $P^y Q^y = I = Q^y P^y$, so

$$(18.7) \qquad\qquad p_2 \# q_{-2} \equiv 1 \equiv q_{-2} \# p_2.$$

We may now follow the classical parametrix construction as outlined in §9. In fact (18.7) and Theorem 14.1 imply

$$(18.8) \qquad\qquad PQ_{-2} = I - R, \quad Q_{-2}P = I - R',$$

where

$$(18.9) \qquad\qquad R, R' \text{ have symbols } r, r' \in S_\nu^{-1}(U).$$

Since $P$ and $Q_{-2}$ are properly supported, so are $R$ and $R'$. Then

$$(18.10) \qquad\qquad PQ_{-2}(I + R + \cdots + R^k) = I - R^{k+1}$$

and $R^{k+1}$ has symbol $r_{k+1} \in S_\nu^{-k-1}(U)$. The term $q_{-2-j}$ in the asymptotic expansion of the symbol of $Q_{-2}(I + R + \cdots + R^k)$ is independent of $k$ for $k \geq j$. Therefore we may choose $q \in S^{-2}(U)$, $q \sim \sum q_{-2-j}$ so that the operator $Q = q(x, D)$ is properly supported and satisfies

$$(18.11) \qquad\qquad PQ - I \in \bigcap_k S_\nu^{-2-k}(U) = S^{-\infty}(U).$$

Considering $(I + R' + \cdots + [R']^k)Q_{-2}$ instead allows us to find $Q'$ with symbol in $S_\nu^{-2}(U)$ such that $Q'P \sim I$. Then

$$(18.12) \qquad\begin{aligned} Q' &= Q'(PQ - [PQ - I]) \sim Q'PQ \\ &= Q + (Q'P - I)Q \sim Q. \end{aligned}$$

Thus $QP \sim I$ and $Q$ is the desired parametrix.

(ii) *Proof that (c)* $\implies$ *(d).* Suppose $Q$ is a parametrix for $P$ having symbol $q \in S_\nu^{-2}(U)$. Again we may assume that $Q$ is properly supported. In view of Proposition 10.22, $q \in S_{\frac{1}{2},\frac{1}{2}}^{-1}(U)$. Then Theorem 9.69 implies

$$(18.13) \qquad\qquad Q : H_{\text{loc}}^s(U) \to H_{\text{loc}}^{s+1}(U).$$

Now suppose $u \in \mathcal{D}'(U)$ and $Pu \in H_{\text{loc}}^s(U)$. Then

$$(18.14) \qquad\qquad u = QPu + (I - QP)u.$$

The first term on the right is in $H_{\text{loc}}^{s+1}(U)$ by (18.13), and the second term is in $C^\infty(U)$. Therefore $P$ is hypoelliptic with loss of one derivative.

(iii) *Proof that (d) $\Longrightarrow$ (a)*. Recall that in Theorem 2.9 we related the condition $\lambda(y) \notin \Lambda_y$ to an *a priori* estimate (2.10). The same argument shows that $\lambda(y) \notin \Lambda_y$ is a consequence of the following estimate:

$$(18.15) \quad \|X_0^y u\| \, \|u\| + \sum_{j=1}^{d} \|X_j^y u\|^2 \leq C \|P^y u\| \, \|u\|, \quad u \in C_c^\infty(\mathbf{R}^{d+1}).$$

In fact (18.15) implies the weaker estimate

$$(18.16) \quad \|X_0^y u\|^2 \leqq C \|P^y u\|^2, \quad u \in C_c^\infty(\mathbf{R}^{d+1}),$$

and the argument following (2.24) shows that $(18.16) \Longrightarrow \lambda(y) \notin \Lambda_y$. Assuming that $P$ is hypoelliptic with a loss of one derivative, we shall derive several inequalities which lead to (18.15).

Let us fix $y$ and use $y$-coordinates, so that $y$ is taken as the origin in $\mathbf{R}^{d+1}$. Let $K$ be a compact neighborhood of $0$, with $K \subset U$. There is a constant $C_0$ such that

$$(18.17) \quad \|u\|_1 \leq C_0(\|Pu\| + \|u\|), \quad u \in C_v^\infty(U), \quad \text{supp } u \subset K,$$

where $\| \ \|$ and $\| \ \|_1$ denote the $L^2$ and $H^1$ norms. To see this, note that the hypoellipticity assumption means that the completion of this space of test functions with respect to the norm $\|Pu\| + \|u\|$ is contained in $H^1$ and (18.7) follows from the closed graph theorem.

Next, there is a constant $C$ such that

$$(18.18) \quad \|X_0 u\| \, \|u\| + \sum_{j=1}^{d} \|X_j u\|^2 \leq C(\|Pu\| + \|u\|)\|u\|,$$
$$u \in C_c^\infty(U), \quad \text{supp } u \subset K.$$

In fact (18.17) implies that the first term on the left in (18.18) is dominated by the expression on the right in (18.18). As for the remaining terms,

$$(18.19) \quad \sum_{j=1}^{d} \|X_j u\|^2 = -\left(\sum_{j=1}^{d} X_j^2 u, u\right) = (Pu, u) - (Lu, u),$$

where $L$ is a first order operator. Now (18.17) implies that $-(Lu, u)$ is dominated by the right side of (18.18), and we have established (18.18).

To pass from (18.18) to (18.15) we use the dilations associated to the $y$-group sructure. Still in the $y$-coordinates, let

$$(18.20) \quad V_t u(x) = t^{\frac{1}{2}(d+2)} u(t^2 x_0, t x_1, \ldots, t x_d), \quad t > 0.$$

This is a unitary operator in $L^2(\mathbf{R}^{d+1})$, and

(18.21) $$V_t^{-1} X_0^y V_t = t^2 X_0^y, \quad V_t^{-1} X_j^y V_t = t X_j^y, \quad j > 0.$$

Thus also

(18.22) $$V_t^{-1} P^y V_t = t^2 P^y.$$

Let us say that $h \in C^\infty(U)$ has *weight* $\geq k$ if

(18.23) $$|h(x)| \leq C(|x_0| + |x'|^2)^{k/2}, \quad x \in U.$$

If $u \in C_c^\infty(\mathbf{R}^{d+1})$, then $V_t u$ has support in $K$ for large $t$. If $h \in C^\infty(U)$ has weight $\geq k$ it is easy to see that

(18.24) $$\|h V_t u\| = O(t^{-k}) \quad \text{as} \quad t \to +\infty.$$

Consequently, if $Q$ is an operator which is homogeneous of degree $j$ with respect to the $y$-dilations, then

(18.25) $$\begin{aligned} \|h Q V_t u\| &= t^j \|h V_t Q u\| \\ &= O(t^{j-k}) \quad \text{as} \quad t \to +\infty. \end{aligned}$$

Now (11.30)–(11.32) imply

(18.26) $$X_j = X_j^y + \sum_{k=0}^{d} \alpha_{jk} X_k^y$$

with

(18.27)  (degree of homogeneity of $X_k^y$ – weight of $\alpha_{jk}$) $\leq 1$, all $j, k$;

(18.28)  (degree of homogeneity of $X_0^y$ – weight of $\alpha_{j0}$) $\leq 0$ if $j > 0$.

Thus

(18.29) $$P = P^y + \{ \text{ sum of terms with (degree – weight)} \leq 1 \}.$$

Suppose $u \in C_c^\infty(\mathbf{R}^{d+1})$. Combining (18.25)–(18.29) and (18.18) we obtain for large $t$

(18.30)

$$\|X_0^y u\| \|u\| + \sum_{j=1}^{d} \|X_j^y u\|^2 = t^{-2}\left\{\|X_o^y V_t u\| \|V_t u\| + \sum_{j=1}^{d}\|X_j^y V_t u\|^2\right\}$$

$$= t^{-2}\left\{\|X_0 V_t u\| \|V_t u\|\right.$$

$$\left. + \sum_{j=1}^{d}\|X_j V_t u\|^2\right\} + O(t^{-1})$$

$$\leq Ct^{-2}(\|PV_t u\| + \|V_t u\|)\|V_t u\| + O(t^{-1})$$

$$= Ct^{-2}\|P^y V_t u\| \|V_t u\| + O(t^{-1})$$

$$= C\|P^y u\| \|V_t u\| + O(t^{-1}).$$

Taking $t \to +\infty$ we obtain (18.15), and the proof is complete.

Finally we have

(18.31) THEOREM: *Let $u \in \mathcal{D}'(U)$ and suppose that $Pu \in H_{loc}^s(U)$, $\lambda(y) \notin \Lambda_y$. Then $u, X_j u, X_k X_l u \in H_{loc}^s(U)$, where $j = 0,1,\ldots,d$ and $k,l = 1,\ldots,d$. In particular $u \in H_{loc}^{s+1}(U)$ and $P$ is hypoelliptic.*

(18.32) REMARK: The proof of Theorem 18.31 follows from applying $X_j$, $j = 0,1,\ldots,d$, to the parametrix $Q$ the required number of times. As a consequence, the closed graph theorem implies the following *a priori* estimate:

(18.33)
$$\|u\|_s + \sum_{j=0}^{d}\|X_j u\|_s$$
$$+ \sum_{k,l=1}^{d}\|X_k X_l u\|_s, \leq C(\|Pu\|_s + \|u\|_s), \quad u \in C_c^\infty(U).$$

(18.34) REMARK: Let $\sigma(Q)_m(z,\xi) = g_m(z, \sigma(z,\xi))$ denote the principal symbol of the $\mathcal{V}$-operator $Q$. Then $Q^y$ has symbol $g_m(y, \sigma^y(x,\xi))$, where $x$ denotes $y$-coordinates. Given

(18.35)
$$P = -\sum_{j=1}^{d} X_j^2 - i\lambda X_0 + \cdots$$

as in (18.1), we have $\sigma(P^y)(x,\xi) = \sigma(P^y)(0, \sigma^y(x,\xi))$, and $\sigma(P^y)(0,\xi) = \sigma(P)_2(0,\xi)$, where $\sigma(P)_2$ denotes the principal symbol of $P$. Using (1.10)

$$(18.36) \qquad \sigma(P)_2(0,\xi) = \sum_{j=1}^{d} \xi_j^2 + \frac{1}{2}i\sum_{j=1}^{d} b_{jj}\xi_0 + \lambda\xi_0,$$

which justifies (18.2). A similar argument shows that given a partial differential operator

$$(18.37) \qquad P(x,X) = P_m(x,X) + \cdots$$

expressed as a $\mathcal{V}$-operator, we have

$$(18.38) \qquad P^y = P_m(y, X^y).$$

## §19 $\mathcal{V}$-Operators on Compact Manifolds – Hilbert Space Theory

In this section we prepare for the $L^2$ theory of $\square_b$ in the next chapter by considering $\mathcal{V}$-operators on compact manifolds. With this application in mind we consider systems, i.e. operators on sections of bundles, rather than restrict to the scalar case. We simply note that the theory of this chapter carries over readily to systems.

Suppose $M$ is a compact manifold with a smooth positive density $dx$, and suppose $E$ is a $C^\infty$-bector bundle over $M$ equipped with a positive definite hermitian inner product $\langle\,,\,\rangle$. Then $C^\infty(M; E)$ carries an inner product

$$(u,v) = \int_M \langle u(x), v(x)\rangle dx$$

and the completion will be denoted $L^2(M; E)$.

Suppose now that $E_1$ and $E_2$ are two such bundles over $M$, and suppose $M$ is equipped with a subbundle $\mathcal{V} \subset TM$ having fibre codimension one.

(19.1) DEFINITION: *An operator* $Q : C^\infty(M; E_1) \to C^\infty(M; E_2)$ *is a $\mathcal{V}$-operator if for each coordinate neighborhood $U \subset M$ which carries trivializations of $E_1$ and $E_2$, each element of the corresponding matrix of operators $Q_{jk} : C_c^\infty(U) \to C^\infty(U)$ is a $\mathcal{V}$-operator. If each $Q_{jk}$ has order $m$, we say that $Q$ has order $m$.*

(19.2) REMARKS: As is the case locally, a $\mathcal{V}$-operator $Q$ has a unique continuous extension mapping $\mathcal{D}'(M; E_1)$ to $\mathcal{D}'(M; E_2)$. We denote this extension also by $Q$. Note that $Q$ has a formal adjoint $Q^*$,

(19.3)     $(Q^*u, v) = (u, Qv)$,   $u \in C^\infty(M; E_2)$,   $v \in C^\infty(M; E_1)$.

By Theorem 17.4 the formal adjoint is again a $\mathcal{V}$-operator.

(19.4) DEFINITION: *If $Q : \mathcal{D}'(M; E_1) \to \mathcal{D}'(M; E_2)$ is a $\mathcal{V}$-operator, its $L^2$-realization is the operator*

$$[Q] : L^2(M; E_1) \to L^2(M; E_2).$$

*which is the restriction of $Q$ to the domain*

$$\mathrm{dom}[Q] = \{u \in L^2(M; E_1) : Qu \in L^2(M; E_2)\}.$$

(19.5) PROPOSITION: *The $L^2$-realization of a $\mathcal{V}$-operator is closed and densely defined.*

Proof. If $Q$ is a $\mathcal{V}$-operator, then $\mathrm{dom}[Q]$ contains the dense subspace $C^\infty(M; E_1)$. If $(u_n)_{n=1}^\infty \subset \mathrm{dom}[Q]$ and $u_n \to u$ in $L^2(M; E_1)$ while $Qu_n \to v$ in $L^2(M; E_2)$, then $u_n \to u$ and $Qu_n \to v$ as distributions. Therefore $Qu = v$ and $u \in \mathrm{dom}[Q]$.

(19.6) PROPOSITION: *If $Q$ is a $\mathcal{V}$-operator of order $\leq 0$, then its $L^2$-realization $[Q]$ is a bounded operator. Moreover, the adjoint $[Q]^*$ is the realization $[Q^*]$ of the formal adjoint.*

Proof. $\mathcal{V}$-operators are of type $(\frac{1}{2}, \frac{1}{2})$, so the boundedness is a consequence of the theorem of Calderon and Vaillancourt, Theorem 9.69. The operator $[Q^*]$ is also bounded. Since (19.3) holds on the dense subspaces $C^\infty(M; E_j)$, it carries over to the completions and $[Q]^* = [Q^*]$.

(19.7) DEFINITION: *Suppose $P$ is a closed operator with domain $\mathrm{dom}\, P \subset H_1$ and range $\mathrm{ran}\, P \subset H^2$, where the $H_j$ are Hilbert spaces. Suppose also that $\mathrm{ran}\, P$ is closed in $H_2$. By the associated projections we mean the orthogonal projections $\Pi_j = \Pi_j(P)$ in $H_j$ such that*

$$\mathrm{ran}\, \Pi_1 = (\ker P)^\perp, \quad \mathrm{ran}\, \Pi_2 = \mathrm{ran}\, P.$$

*By the partial inverse of $P$ we mean the unique bounded operator $A : H_2 \to H_1$ with*

$$PA = \Pi_2 \text{ on } H_2, \quad AP = \Pi_1 \text{ on } \mathrm{dom}\, P.$$

The following two theorems will be very useful in studying Hilbert space properties of $\Box_b$.

(19.8) THEOREM: *Suppose $P$, $\Pi_1$, $\Pi_2$, and $A$ are $\mathcal{V}$-operators with*

$$P : \mathcal{D}'(M; E_1) \to \mathcal{D}'(M; E_2)$$
$$\Pi_j : \mathcal{D}'(M; E_j) \to \mathcal{D}'(M; E_j), \quad j = 1, 2,$$
$$A : \mathcal{D}'(M; E_2) \to \mathcal{D}'(M' E_1).$$

*Suppose that the $\Pi_j$ and $A$ are of order $\leq 0$ and that*

(19.9)                         $$PA = \Pi_2, \quad AP = \Pi_1,$$

(19.10)                        $$\Pi_2 P = P = P\Pi_1.$$

(19.11)                        $$\Pi_j^2 = \Pi_j = \Pi_j^*.$$

*Then the $L^2$-realization $[P]$ has closed range, $[\Pi_1]$ and $[\Pi_2]$ are the associated projections and $[A]$ is the partial inverse. Moreover, $[P]$ is the closure of the restriction of $P$ to $C^\infty(M; E_1)$ and $[P]^* = [P^*]$, the $L^2$-realization of the formal adjoint.*

Proof. By (19.11) and Proposition 19.6 the $[\Pi_j]$ are orthogonal projections. Moreover, $[A]$ is a bounded operator. We need to show

(19.12)              $$\operatorname{ran}[P] = \operatorname{ran}[\Pi_2], \quad \ker[P] = \ker[\Pi_1].$$

Now $v \in \operatorname{ran}[\Pi_2]$ implies $v = \Pi_2 v = PAv$ and $u = Av$ is in $L^2(M; E_1)$. Therefore $v = Pu \in \operatorname{ran}[P]$. Conversely suppose $u \in L^2(M; E_1)$ and $Pu \in L^2(M; E_2)$. Then $Pu = \Pi_2 Pu$ is in $\operatorname{ran}[\Pi_2]$. To prove the second part of (19.12) we note that if $u \in L^2(M; E_1)$ then $Pu = P\Pi_1 u$ so $\ker[P] \supset \ker[\Pi_1]$, while $\Pi_1 u = APu$ so $\ker[P] \subset \ker[\Pi_1]$.

To show that $[P]$ is the closure of $P$ on $C^\infty(M; E_1)$, we assume $u \in \operatorname{dom}[P]$. Choose sequences $(u_n)_{n=1}^\infty \subset \mathcal{V}^\infty(M; E_1)$ such that $u_n \to u$ in $L^2(M; E_1)$ and $(v_n)_{n=1}^\infty \subset C^\infty(M; E_2)$ such that $u_n \to v = Pu$ in $L^2(M; E_2)$. Then $u_n' = (I - \Pi_1)u_n + Av_n$ is in $C^\infty(M; E_1)$ and

$$u_n' \to (I - \Pi_1)u + APu = (I - \Pi_1)u + \Pi_1 u = u$$

in $L^2(M; E_1)$, while

$$Pu_n' = PAv_n = \Pi_2 v_n \to \Pi_2 Pu = Pu$$

in $L^2(M; E_2)$.

Finally suppose $u \in \operatorname{dom}[P]^*$ and $[P]^* u = v$. Considering $u$ and $v$ as distributions, we have

(19.13)           $$(P^* u, \phi) = (u, P\phi) = (v, \phi), \quad \phi \in C^\infty(M; E_1).$$

Therefore $P^*u = v$ and $u$ belongs to dom$[P^*]$. Conversely, suppose $u \in L^2(M; E_2)$ and $P^*u = v \in L^2(M; E_1)$. Then (19.13) holds, so $u$ is in the domain of the adjoint of the restriction of $P$ to $C^\infty(M; E_1)$. We have shown that the closure of this restriction is $[P]$, so we have $u \in$ dom$[P]^*$ and $[P]^*u = v$.

To complete our general discussion of the $L^2$ theory, we consider parametrices.

(19.14) REMARK: Suppose $P : C^\infty(M; E_1) \to C^\infty(M; E_2)$ is a $\mathcal{V}$-operator and suppose the restriction to $C_c^\infty(U)$ has a two-sided parametrix $Q_U$ which is a $\mathcal{V}$-operator, for each member of a cover $\{U\}$ of $M$ by coordinate charts. Then $P$ has a global parametrix $Q$ which is a $\mathcal{V}$-operator. In fact we may assume $\{U\}$ is a finite cover and that each $Q_U$ is properly supported. Let $\{\phi_U\}$ be a $C^\infty$-partition of unity subordinate to $\{U\}$ and set

$$(19.15) \qquad Qv = \sum_U Q_U(\phi_U v), \quad v \in C^\infty(M; E_2).$$

(19.16) THEOREM: *Suppose $P$ is a $\mathcal{V}$-operator with parametrix $Q$ and suppose $Q$ is a $\mathcal{V}$-operator of degree $\leq 0$. Then the $L^2$-realization $[P]$ has the following properties:*

$$(19.17) \qquad \dim \ker[P] < \infty \quad and \quad \ker[P] \subset C^\infty(M; E_1).$$

$$(19.18) \qquad \text{ran}[P] \text{ is closed and } \text{codim ran}[P] < \infty.$$

$$(19.19) \qquad \begin{array}{l} \textit{The associated projections and the partial inverse} \\ \textit{of } [P] \textit{ are } \mathcal{V}\textit{-operators.} \end{array}$$

Proof. We have

$$(19.20) \qquad QP = I - S_1, \quad PQ = I - S_2$$

where the $S_j$ are smoothing operators. Therefore $\ker[P] \subset \ker[I - S_1]$. Now $[I - S_1] = I - [S_1]$ and $[S_1]$ is compact, so $\ker[I - S_1]$ is finite-dimensional. Since $S_1 : L^2(M; E_1) \to C^\infty(M; E_1)$, $\ker[I - S_1] \subset C^\infty(M; E_1)$. Similarly, $[Q]$ is bounded so $\text{ran}[P] \supset \text{ran}[I - S_2] = \text{ran}(I - [S_2])$ and the latter range is closed and has finite codimension.

Let $\{\phi_1, \dots, \phi_m\}$ be an orthonormal basis for $\ker[P]$. The projection $\Pi_0 = I - \Pi_1$ is given by

$$(19.21) \qquad \Pi_0 u = \sum_{j=1}^m (u, \phi_j)\phi_j.$$

Since $\phi_j \in C^\infty(M; E_1)$, $\Pi_0$ extends to map $\mathcal{D}'(M; E_1)$ to $C^\infty(M; E_1)$. Thus $\Pi_1$ differs from the identity by a smoothing operator, so $\Pi_1$ is a $\mathcal{V}$-operator. Note that $(\mathrm{ran}[P])^\perp \subset \ker P^*$ and $P^*$ has parametrix $Q^*$, so $(\mathrm{ran}[P])^\perp \subset C^\infty(M; E_2)$ and $\Pi_0' = I - \Pi_2$ is also a smoothing operator and $\Pi_2$ is a $\mathcal{V}$-operator.

Finally, let $A$ be the partial inverse of $[P]$. Then

$$
\begin{aligned}
A - Q &= (QP + S_1)A - Q = Q\Pi_2 + S_1 A - Q \\
&= -Q\Pi_0' + S_1 A = -Q\Pi_0' + S_1 A(PQ + S_2) \\
&= -Q\Pi_0' + S_1 \Pi_1 Q + S_1 A S_2.
\end{aligned}
$$

The first two terms on the right in the last expression are smoothing operators, since $S_1$ and $\Pi_0'$ are smoothing. Also, $S_1 A S_2 : \mathcal{D}'(M; E_2) \to L^2(M; E_2) \to L^2(M; E_1) \to C^\infty(M; E_1)$. Thus $A - Q$ is smoothing and $A$ is a $\mathcal{V}$-operator.

# CHAPTER 4

# Applications to the $\overline{\partial}_b$-Complex

## §20 The $\overline{\partial}_b$-Complex and $\square_b$

Let $M$ denote a $C^\infty$-manifold of dimension $2n + 1$ equipped with a *Cauchy–Riemann or CR-structure*, i.e. a complex rank $n$ sub-bundle, $T_{1,0}$, of the complexified tangent bundle $CTM$, which has the following properties

$$(20.1) \qquad T_{1,0} \cap T_{0,1} = \{0\}, \text{ where we set } T_{0,1} = \overline{T_{1,0}},$$

$$(20.2) \qquad \begin{array}{l} T_{1,0} \text{ (and therefore } T_{0,1}) \text{ is integrable in Frobenius' sense :} \\ \text{if } Z \text{ and } W \text{ are sections of } T_{1,0}, \text{ then so is } [Z, W]. \end{array}$$

We assume that $M$ is equipped with a smoothly varying positive definite Hermitian form on $CTM$, which is compatible with the $CR$-structure:

$$(20.3) \quad T_{1,0} \perp T_{0,1} \text{ and complex conjugation is an isometry in } CTM.$$

Then there is a unique real line bundle $N \subset TM$ such that $N \perp T_{1,0} \oplus T_{0,1}$. In particular we have the orthogonal decomposition:

$$(20.4) \qquad\qquad CTM = T_{1,0} \oplus T_{0,1} \oplus CN.$$

The $\overline{\partial}_b$-complex of Kohn and Rossi [1] may be realized as follows. We define the bundle of covectors, $\Lambda^{1,0}$, of type $(1,0)$ by

$$(20.5) \qquad\qquad \Lambda^{1,0} = (T_{0,1} \oplus N)^\perp \subset CT^*M.$$

Similarly, we set

$$(20.6) \qquad\qquad \Lambda^{0,1} = (T_{1,0} \oplus N)^\perp \subset CT^*M.$$

The bundle of covectors of type $(0, q)$ is

$$(20.7) \qquad\qquad \Lambda^{0,q} = (\Lambda^{0,1})^q \subset \Lambda^q = \Lambda^q(CT^*M),$$

where the exponents refer to the iterated wedge product. The Hermitian form induces, by duality, an inner product on each fiber of $CT^*M$, and therefore an inner product, $\langle\,,\,\rangle$, on each fiber of $\Lambda^q$. Let $\Pi_q$ denote the natural projection

(20.8)                                 $\Pi_q : \Lambda^q \to \Lambda^{0,q}.$

A $(0,q)$-form is a section of $\Lambda^{0,q}$. For a fixed $q$

(20.9)                                 $\bar\partial_b : \Lambda^{0,q} \to \Lambda^{0,q+1}$

is defined by

(20.10)                                $\bar\partial_b = \Pi_{q+1} \circ d,$

where $d$ denotes the usual exterior derivative.

(20.11) PROPOSITION:

$$\bar\partial_b^2 = 0.$$

Proof. If $X, Y \in T_{0,1}$ and $\omega$ is a 1-form in $(T_{0,1})^\perp$, then (20.1) and (20.2) imply

(20.12)        $d\omega(X,Y) = X(\omega(Y)) - Y(\omega(X)) - \omega([X,y]) = 0.$

If $f \in \Lambda^{0,q}$, then

$$0 = d^2 f = d(\Pi df + (I - \Pi)df) = d(\Pi df) + dg.$$

Now $g = df - \Pi df$ is in the ideal generated by the 1-forms which annihilate $T_{0,1}$ and (20.12) shows that this ideal is $d$-closed, hence $\Pi dg = 0$. Consequently

$$0 = \Pi d^2 f = \Pi d(\Pi df) + \Pi dg = \bar\partial_b^2 f.$$

The pointwise inner product on $\Lambda^{0,q}$ induces an inner product on sections of $\Lambda^{0,q}$:

$$(f,g) = \int_M \langle f(x), g(x)\rangle dV(x),$$

where $f, g \in C_c^\infty(M, \Lambda^{0,q})$ and $dV$ is the volume form. Then $\bar\partial_b$ has a formal adjoint $\vartheta_b$,

(20.13)                                $\vartheta_b : \Lambda^{0,q+1} \to \Lambda^{0,q}.$

We shall compute it presently. The associated Laplacian, $\Box_b$,

(20.14)                                $\Box_b : \Lambda^{0,q} \to \Lambda^{0,q}$

is given by

(20.15)                                $\Box_b = \vartheta_b \bar\partial_b + \bar\partial_b \vartheta_b.$

In the rest of §20 we evaluate $\square_b$. Locally we may choose an orthonormal frame $\omega^1, \omega^2, \ldots, \omega^n$ for $\Lambda^{1,0}$; then $\overline{\omega}^1, \overline{\omega}^2, \ldots, \overline{\omega}^n$ is an orthonormal frame for $\Lambda^{0,1}$. The $2n$-form

$$(20.16) \qquad \omega = i^n \omega^1 \wedge \overline{\omega}^1 \wedge \cdots \wedge \omega^n \wedge \overline{\omega}^n$$

is real and is independent of the choice of the orthonormal frame; thus $\omega$ may be considered as a globally defined element of $\Lambda^{n,n}$. Again, locally there is a real 1-form $\omega^0$ of length one which is orthogonal to $\Lambda^{1,0} \oplus \Lambda^{0,1}$. $\omega^0$ is unique up to sign and can be specified uniquely by requiring that the map

$$(20.17) \qquad f \to \int_m f \omega^0 \wedge \omega, \quad f \in C_c(U),$$

define a positive measure on the domain of definition of $\omega^0$. Therefore $\omega^0$, so chosen, is a uniquely determined global 1-form, and

$$(20.18) \qquad dV = \omega^0 \wedge \omega$$

may be taken to be the volume form on $M$. Note that $\omega^0$ annihilates $T_{1,0} \oplus T_{0,1}$.

(20.19) DEFINITION: *The Levi form is the Hermitian function-valued quadratic form defined on sections of $T_{1,0}$:*

$$(20.20) \qquad \mathcal{L}(Z, W) = i\omega^0([Z, \overline{W}]), \quad Z, W \in T_{1,0},$$

*where $\omega^0$ is the annihilator of $T_{1,0} \oplus T_{0,1}$ chosen above.*

Note that the usual derivation formula yields

$$(20.21) \qquad \mathcal{L}(Z, W) = -id\omega^0(Z, \overline{W}).$$

Therefore $\mathcal{L}$ induces pointwise (or is induced by) the form $-id\omega_x^0$ on $(T_{1,0})_x$.

(20.22) DEFINITION: *The eigenvalues of the Levi form at $y \in M$ are the eigenvalues of the Hermitian form $\mathcal{L}^y = -id\omega^0(y)$ with respect to the inner product $\langle\,,\,\rangle$ on $(T)_{1,0})_y$.*

Next we compute $\overline{\partial}_b$ and $\vartheta_b$. Let $Z_j$, $j = 1, \ldots, n$ denote the basis of $T_{1,0}$ which is dual to $\omega^j$, $j = 1, \ldots, n$, i.e.

$$(20.23) \qquad \langle df, \omega^j \rangle = Z_j f, \quad j = 1, \ldots, n, \quad f \in C_c^\infty(M).$$

Then

$$(20.24) \qquad df = X_0(f)\omega^0 + \sum_{j=1}^n Z_j(f)\omega^j + \sum_{j=1}^n \overline{Z}_j(f)\overline{\omega}^j,$$

which defines $X_0$, and

$$(20.25) \qquad \overline{\partial}_b f = \sum_{j=1}^{n} \overline{Z}_j(f) \overline{\omega}^j, \quad f \in C_c^\infty(M).$$

In general, suppose $f \in \Lambda^{0,q}$, i.e.

$$(20.26) \qquad f = \sum_{|J|=q} f_J \overline{\omega}^J,$$

where $|J|$ denotes the length of the multi-index $J = (j_1, \ldots, j_q)$, $1 \leq j_1 < \cdots < j_q \leq n$, and

$$(20.27) \qquad \{\overline{\omega}^J = \overline{\omega}^{j_1} \wedge \cdots \wedge \overline{\omega}^{j_q}, \quad 1 \leq j_1 < \cdots < j_q \leq n\}$$

is the induced orthonormal basis of $\Lambda^{0,q}$. Then

$$(20.28) \qquad df = \sum_{|J|=q} (df_J) \wedge \overline{\omega}^J + \varepsilon(f),$$

where $\varepsilon(f)$ denotes a multilinear map of $\{f_J, |J| = q\}$ which does not involve derivatives of $f_J$. Projecting onto $(0, q+1)$-forms we find

$$(20.29) \qquad \overline{\partial}_b f = \sum_{|J|=q} \sum_{j=1}^{n} \overline{Z}_j(f_J) \overline{\omega}^j \wedge \overline{\omega}^J + \varepsilon(f).$$

As for $\vartheta_b$, let $f \in \Lambda^{0,q+1}$ and $q \in \Lambda^{0,q}$. Then

$$(20.30) \qquad \begin{aligned} (\vartheta_b f, g) &= (f, \overline{\partial}_b g) = \int \langle f, \overline{\partial}_b g \rangle dV \\ &= \sum_{|K|=q+1} \int f_K \sum_{(j,J)\sim K} \varepsilon(j, J) \overline{\overline{Z}_j(g_J)} dV + \varepsilon(f, g), \end{aligned}$$

where $(j, J) \sim K$ means that there is a permutation taking the ordered set $(j, J) = (j, j_1, \ldots, j_q)$ to $(k_1, \ldots, k_q, k_{q+1}) = K$ and $\varepsilon(j, J)$ is the sign of that permutation. If $j \in J$ we set $\varepsilon(j, J) = 0$. Integration by parts yields

$$(20.31) \qquad \vartheta_b f = - \sum_{|J|=q} \sum_{(j,J)\sim K} \varepsilon(j, J) Z_j(f_K) \overline{\omega}^J + \varepsilon(f).$$

This can be expressed more simply in terms of the interior product:

$$(20.32) \qquad \overline{\omega}^j \lrcorner \overline{\omega}^K = \begin{cases} 0 & \text{if } j \in \{k_1, \ldots, k_{q+1}\}, \\ \varepsilon(j, J) \overline{\omega}^J & \text{if } (j, J) \sim K. \end{cases}$$

Thus, for $f \in \Lambda^{0,q+1}$

$$(20.33) \qquad \vartheta_b f = - \sum_{|K|=q+1} \sum_{j=1}^{n} Z_j(f_K) \overline{\omega}^j \lrcorner \overline{\omega}^K + \varepsilon(f).$$

Now we are ready to compute $\square_b$. Thus

$$\vartheta_b \overline{\partial}_b \left( \sum_{|J|=q} f_J \overline{\omega}^J \right) = \vartheta_b \left( \sum_J \sum_{j=1}^n \overline{Z}_j(f_J) \overline{\omega}^j \wedge \overline{\omega}^J + \varepsilon(f) \right)$$

(20.34)

$$= -\sum_J \sum_{i,j=1}^n Z_i \overline{Z}_j(f_J) \overline{\omega}^i \lrcorner (\overline{\omega}^j \wedge \overline{\omega}^J)$$

$$+ \varepsilon(f, Zf, \overline{Z}f),$$

where $Zf$ denotes the collection $\{Z_j(f_J)\}$, and similarly for $\overline{Z}f$. Also

$$\overline{\partial}\vartheta_b \left( \sum_{|J|=q} f_J \overline{\omega}^J \right) = \overline{\partial}_b \left( -\sum_J \sum_{i=1}^n Z_i(f_J)\overline{\omega}^i \lrcorner \overline{\omega}^J \right) + \varepsilon(f)$$

(20.35)

$$= -\sum_J \sum_{i,j=1}^n \overline{Z}_j Z_i(f_J) \overline{\omega}^j \wedge (\overline{\omega}^i \lrcorner \overline{\omega}^J)$$

$$+ \varepsilon(f, Zf, \overline{Z}f).$$

Consequently we have

$$\square_b \left( \sum_{|J|=q} f_J \overline{\omega}^J \right) = -\sum_J \left\{ \sum_{i,j=1}^n Z_i \overline{Z}_j (f_J) \overline{\omega}^i \lrcorner (\overline{\omega}^j \wedge \overline{\omega}^J) \right.$$

$$\left. + \sum_{i,j=1}^n \overline{Z}_j Z_i (f_J) \overline{\omega}^j \wedge (\overline{\omega}^i \lrcorner \overline{\omega}^J) \right\}$$

$$+ \varepsilon(f, Zf, \overline{Z}f)$$

$$= -\sum_J \left\{ \sum_{j=1}^n Z_j \overline{Z}_j (f_J) \overline{\omega}^j \lrcorner (\overline{\omega}^j \wedge \overline{\omega}^J) \right.$$

(20.36)

$$\left. + \sum_{j=1}^n \overline{Z}_j Z_j (f_J) \overline{\omega}^j \wedge \overline{\omega}^j \lrcorner \overline{\omega}^J) \right\}$$

$$- \sum_J \left\{ \sum_{i \neq j} Z_i \overline{Z}_j (f_J) \overline{\omega}^i \lrcorner (\overline{\omega}^j \wedge \overline{\omega}^J) \right.$$

$$\left. + \sum_{i \neq j} \overline{Z}_j Z_i (f_J) \overline{\omega}^j \wedge (\overline{\omega}^i \lrcorner \overline{\omega}^J) \right\}$$

$$+ \varepsilon(f, Zf, \overline{Z}f).$$

To simplify (20.36) we note that

(20.37)
$$\overline{\omega}^j \lrcorner (\overline{\omega}^j \wedge \overline{\omega}^J) = \begin{cases} 0 & \text{if } j \in \{J\}, \\ \overline{\omega}^J & \text{if } j \notin \{J\}, \end{cases}$$

(20.38) $$\overline{\omega}^j \wedge (\overline{\omega}^j \lrcorner \overline{\omega}^J) = \begin{cases} \overline{\omega}^J & \text{if } j \in \{J\}, \\ 0 & \text{if } j \notin \{J\}, \end{cases}$$

and

(20.39) $$\overline{\omega}^j \wedge (\overline{\omega}^i \lrcorner \overline{\omega}^J) = -\overline{\omega}^i \lrcorner (\overline{\omega}^j \wedge \overline{\omega}^J) \quad \text{if } i \neq j.$$

Consequently

(20.40) $$\Box_b \left( \sum_{|J|=q} f_J \overline{\omega}^J \right) = -\sum_J \left\{ \left( \sum_{j \notin \{J\}} Z_j \overline{Z}_j + \sum_{j \in \{J\}} \overline{Z}_j Z_j \right) (f_J) \right\} \overline{\omega}^J$$
$$- \sum_J \sum_{i \neq j} [Z_i, \overline{Z}_j](f_J) \overline{\omega}^i \lrcorner (\overline{\omega}^j \wedge \overline{\omega}^J) + \varepsilon(f, Zf, \overline{Z}f).$$

Finally we write

(20.41) $$\sum_{j \notin \{J\}} Z_j \overline{Z}_j + \sum_{j \in \{J\}} \overline{Z}_j Z_j = \frac{1}{2} \sum_{j=1} (Z_j \overline{Z}_j + \overline{Z}_j Z_j)$$
$$+ \frac{1}{2} \sum_{j \notin \{J\}} [Z_j, \overline{Z}_j] - \frac{1}{2} \sum_{j \in \{J\}} [Z_j, \overline{Z}_j].$$

Thus we have derived

(20.42) PROPOSITION: *On $(0,q)$-forms $\Box_b$ is given by*

(20.43) $$\Box_b \left( \sum_{|J|=q} f_J \overline{\omega}^J \right) = -\sum_J \left\{ \frac{1}{2} \sum_{j=1}^{n} (Z_j \overline{Z}_j + \overline{Z}_j Z_j) + \frac{1}{2} \sum_{j \notin \{J\}} [Z_j, \overline{Z}_j] \right.$$
$$\left. - \frac{1}{2} \sum_{j \in \{J\}} [Z_j, \overline{Z}_j] \right\} (f_J) \overline{\omega}^J$$
$$- \sum_J \sum_{i \neq j} [Z_i, \overline{Z}_j](f_J) \overline{\omega}^i \lrcorner (\overline{\omega}^j \wedge \overline{\omega}^J)$$
$$+ \varepsilon(f, Zf, \overline{Z}f).$$

§21 *Hypoellipticity of $\Box_b$. Condition $Y(q)$*

From now on $\Box_{b,q}$ will denote $\Box_b$ acting on $(0,q)$-forms. Considering $\Box_{b,q}$ as a $\mathcal{V}$-operator we shall apply the theory of Chapter 3 to find local conditions which guarantee that $\Box_{b,q}$ is hypoelliptic. This requires the careful study of *the y-invariant $\Box_{b,q}^y$*.

In the rest of Chapter 4 we shall use superscripts on coordinates; this agrees with the usual superscript notation for differential forms.

According to Remark 18.34 $\square_{b,q}^y$ is obtained in the following manner. Let $x$ denote the $y$-coordinates for $X_0, X_1, \ldots, X_{2n}$, where

$$(21.1) \qquad Z_j = \frac{1}{2}(X_j - iX_{j+n}), \quad j = 1, \ldots, n.$$

Then

$$(21.2) \qquad X_0 = \frac{\partial}{\partial x^0} + \sum_{k=0}^{2n+1} \beta_{0k}(x)\frac{\partial}{\partial x^k},$$

$$(21.3) \qquad Z_j = \frac{1}{2}\left(\frac{\partial}{\partial x_j} - i\frac{\partial}{\partial x^{j+n}}\right) + \sum_{k=0}^{2n+1} \beta_{jk}(x)\frac{\partial}{\partial x^k}, \quad j = 1, 2, \ldots, n.$$

Therefore the $y$-invariant vector fields are

$$(21.4) \qquad X_0^y = \frac{\partial}{\partial x^0},$$

$$(21.5) \qquad Z_j^y = \frac{\partial}{\partial z_j} + \sum_{k=1}^{n}(a_{jk}z^k + b_{jk}\bar{z}^k)\frac{\partial}{\partial x^0}, \quad j = 1, 2, \ldots, n,$$

where

$$(21.6) \qquad a_{jk} = \frac{\partial \beta_{j0}}{\partial z^k}\bigg|_{x=0}, \quad b_{jk} = \frac{\partial \beta_{j0}}{\partial \bar{z}^k}\bigg|_{x=0}$$

and $z^j = x^j + ix^{j+n}$, $j = 1, 2, \ldots, n$. The integrability of $T_{1,0}$, i.e. $[Z_j, Z_k] \in T_{1,0}$ implies that

$$(21.7) \qquad a_{jk} = a_{kj}, \quad j, k = 1, 2, \ldots, n.$$

Consequently, $\{Z_j^y, j = 1, 2, \ldots, n\}$ generates an integrable bundle $T_{1,0}^y$. We set

$$(21.8) \qquad \omega^{y,0} = dt^0 - \text{Re} \sum_{j,k=1}^{n}(a_{jk}z^k + b_{jk}\bar{z}^k)dz^j,$$

$$(21.9) \qquad \omega^{y,j} = dz^j, \quad j = 1, 2, \ldots, n.$$

Then

$$\Box_{b,q}^y\left(\sum_{|J|=q} f_J d\bar{z}^J\right) = -\sum_J \left\{\frac{1}{2}\sum_{j=1}^n (Z_j^y \overline{Z}_j^y + \overline{Z}_j^y Z_j^y)\right.$$

$$+ \frac{1}{2}\sum_{j\notin\{J\}} [Z_j^y, \overline{Z}_j^y]$$

(21.10)

$$\left. - \frac{1}{2}\sum_{j\in\{J\}} [Z_j^y, \overline{Z}_j^y]\right\}(f_J)d\bar{z}^J$$

$$- \sum_J \sum_{i\neq j} [Z_i^y, \overline{Z}_j^y](f_J)d\bar{z}^i \lrcorner (d\bar{z}^j \wedge d\bar{z}^J).$$

In other words, $\Box_{b,q}^y$ is simply $\Box_{b,q}$ with respect to $T_{1,0}^y$. We note that

(21.11) $$\mathcal{L}_{jk}^y = i\omega^{y,0}([Z_j^y, \overline{Z}_k^y]) = i\omega^0([Z_j, \overline{Z}_k])|_y$$

denotes the Levi form of $T_{1,0}^y$.

$\Box_{b,q}^y$ *in skew-symmetric form.* We rewrite (21.10). Note that

(21.12) $$[Z_j^y, \overline{Z}_k^y] = (\bar{b}_{kj} - b_{jk})\frac{\partial}{\partial x^0} = -2c_{jk}\frac{\partial}{\partial x^0}.$$

Hence

$$\Box_{b,q}^y\left(\sum_{|J|=q} f_J d\bar{z}^J\right) = -\sum_J \left\{\frac{1}{2}\sum_{j=1}^n (Z_j^y \overline{Z}_j^y + \overline{Z}_j^y Z_j^y)\right.$$

(21.13)

$$\left. - \left(\sum_{j\notin\{J\}} c_{jj} - \sum_{j\in\{J\}} c_{jj}\right)\frac{\partial}{\partial x^0}\right\}(f_J)d\bar{z}^J$$

$$+ 2\sum_J \sum_{j\notin k} c_{jk}\frac{\partial}{\partial x^0}(f_J)d\bar{z}^j \lrcorner (d\bar{z}^k \wedge d\bar{z}^J).$$

Then the quadratic coordinate change

(21.14)
$$\begin{cases} v^0 = x^0 - \mathrm{Re}\sum_{j,k=1}^n a_{jk}z^j z^k - \frac{1}{2}\sum_{j,k=1}^n (b_{jk} + \bar{b}_{kj})z^j\bar{z}^k, \\ z^j = v^j + iv^{j+n} = x^j + ix^{j+n}, \quad j = 1,\ldots,n, \end{cases}$$

reduces our vector fields to skew-symmetric (skew-hermitian may be more precise) form:

(21.15)
$$\begin{cases} X_0^y = \dfrac{\partial}{\partial v^0}, \\ Z_j^y = \dfrac{\partial}{\partial z_j} + \displaystyle\sum_{k=1}^n c_{jk}\bar{z}^k\dfrac{\partial}{\partial v^0}, \quad j = 1,\ldots,n. \end{cases}$$

We note that the matrix $-iC = -i(c_{jk})$ is hermitian, i.e. $\overline{(-iC)}^t = -iC$, and

$$(21.16) \qquad \mathcal{L}^y = -2iC.$$

Finally, in skew-symmetric coordinates we also have

$$(21.17) \qquad \begin{cases} \omega^{y,0} = dv^0 - Re \sum_{j,k=1}^{n} c_{jk} \bar{z}^k dz^j, \\ \omega^{y,j} = dz^j, \quad j = 1, \ldots, n. \end{cases}$$

*The normal form of* $\square_{b,q}^y$. The definition of $\square_{b,q}$ is independent of the choice of the orthonormal basis, $Z_1, \ldots, Z_n$, of $T_{1,0}$. Let

$$(21.18) \qquad W_j = \sum_{k=1}^{n} u_{jk} Z_k^y, \quad j = 1, 2, \ldots, n$$

denote another orthonormal basis of $T_{1,0}^y$, with respect to which the matrix

$$(21.19) \qquad i\omega^{y,0}([W_j, \overline{W}_k]) = (U\mathcal{L}^y U^*)_{jk}$$

is diagonal, i.e.

$$(21.20) \qquad U\mathcal{L}^y U^* = \mathrm{diag}(\lambda_1^y, \ldots, \lambda_n^y),$$

where $U = (u_{jk})$. To simplify matters we change coordinates:

$$(21.21) \qquad \begin{cases} u^0 = v^0, \\ w = \overline{U}z, \quad w = (w^1, \ldots, w^n), \quad z = (z^1, \ldots, z^n), \end{cases}$$

where $z^j = v^j + iv^{j+n}$, $j = 1, \ldots, n$. This yields

$$(21.22) \qquad dw = \overline{U}dz,$$

and

$$(21.23) \qquad W_j = \frac{\partial}{\partial w^j} + i\lambda_j^y \overline{w}_j \frac{\partial}{\partial u^0}, \quad j = 1, 2, \ldots, n.$$

In particular $\square_{b,q}^y$ is diagonal:

$$(21.24) \qquad \square_{b,q}^y \left( \sum_{|J|=q} f_J d\overline{w}^J \right) = -\sum_J \left\{ \frac{1}{2} \sum_{j=1}^{n} (W_j \overline{W}_j + \overline{W}_j W_j) \right.$$
$$+ \frac{1}{2} \sum_{j \notin \{J\}} [W_j, \overline{W}_j]$$
$$\left. - \frac{1}{2} \sum_{j \in \{J\}} [W_j, \overline{W}_j] \right\} (f_J) d\overline{w}^J.$$

We set

(21.25) $$w^j = u^j + iu^{j+n}, \quad j = 1, 2, \ldots, n,$$

and

(21.26) $$W_j = \frac{1}{2}(U_j - iU_{j+n}), \quad j = 1, 2, \ldots, n.$$

Then

(21.27) $$U_j = \frac{\partial}{\partial u^j} + 2\lambda_j^y u^{j+n} \frac{\partial}{\partial u^0},$$

(21.28) $$U_{j+n} = \frac{\partial}{\partial u^{j+n}} - 2\lambda_j^y u^j \frac{\partial}{\partial u^0},$$

$j = 1, 2, \ldots, n.$ Consequently

(21.29) $$W_j \overline{W}_j + \overline{W}_j W_j = \frac{1}{2}(U_j^2 + U_{j+n}^2),$$

and

(21.30) $$[W_j, \overline{W}_k] = -2i\lambda_j^y \delta_{jk} \frac{\partial}{\partial u^0}.$$

Therefore

(21.31)
$$\Box_{b,q}^y \left( \sum_{|J|=q} f_J d\overline{w}^J \right) = \frac{1}{4} \sum_J \left\{ \sum_{j=1}^{2n} U_j^2 - i\left( \sum_{j \in \{J\}} 4\lambda_j^y \right. \right.$$
$$\left. \left. - \sum_{j \notin \{J\}} 4\lambda_j^y \right) \frac{\partial}{\partial u^0} \right\} (f_J) d\overline{w}^J$$
$$= \frac{1}{4} \sum_{|J|=q} \Box_J^y (f_J) d\overline{w}^J.$$

In the notation of Definition 2.4

(21.32) $$a_j = -4\lambda_j^y, \quad j = 1, 2, \ldots, n.$$

According to Theorem 2.10, $\Box_J^y$ is invertible if and only if

(21.33) $$\left| \sum_{j \in \{J\}} \lambda_j^y - \sum_{j \notin \{J\}} \lambda_j^y \right| < \sum_{j=1}^n |\lambda_j^y|.$$

(21.34) DEFINITION: *Given $q$, $0 \le q \le n$, the Levi form is said to satisfy condition $Y(q)$ at $y \in M$ if the set of eigenvalues of $\mathcal{L}^y$ cannot be converted to a set of reals all having the same sign by changing exactly $q$ signs.*

Applying Theorem 18.4 to each $\square_J$, $|J| = q$ in the diagonalized $\square_{b,q}^y$ yields the main result of §21:

(21.35) THEOREM: $\square_{b,q}$ has a parametrix in $S_{\mathcal{V}}^{-2}(M)$ if and only if the Levi form satisfies condition $Y(q)$ at each point of $M$, where $\mathcal{V}$ is the bundle $\mathrm{Re}(T_{1,0}) + \mathrm{Im}(T_{1,0})$.

## §22 Parametrix for $\square_{b,q}$

Assuming condition $Y(q)$ at $y \in M$ we give here an explicit construction of the kernel of $(\square_{b,q}^y)^{-1}$. This yields the principal term in the asymptotic expansion of the kernel of $\square_{b,q}^{-1}$. Then, of course, the classical parametrix construction, as outlined in §9, produces the full parametrix for $\square_{b,q}$.

We start with $\square_{b,q}^y$ in normal form and work our way up to the model $\square_{b,q}^y$. Thus, in normal form

$$(22.1) \qquad \square_{b,q}^y \left( \sum_{|J|=q} f_J \, d\bar{w}^J \right) = \frac{1}{4} \sum_{|J|=q} \square_J^y(f_J) \, d\bar{w}^J,$$

where

$$(22.2) \qquad \square_J^y = -\sum_{j=1}^{2n} U_j^2 - i\left( \sum_{j \in \{J\}} 4\lambda_j^y - \sum_{j \notin \{J\}} 4\lambda_j^y \right) \frac{\partial}{\partial u^0},$$

$w^j = u^j + iu^{j+n}$, $j = 1, \ldots, n$ and $U_j$, $j = 1, \ldots, 2n$ is defined by (21.27) and (21.28). $(\square_J^y)^{-1}$ had already been constructed in Theorem 5.38 and we only need to translate it to the present setup. We set

$$(22.3) \qquad A^y(s) = (4\pi)^{-n} \det\left( \frac{4|\mathcal{L}^y|s}{\sinh(4|\mathcal{L}^y|s)} \right)$$

and

$$(22.4) \qquad \gamma^y(s, u') = \frac{1}{4}\left\langle \frac{4|\mathcal{L}^y|s}{\tanh(4|\mathcal{L}^y|s)} w, w \right\rangle,$$

where

$$(22.5) \qquad |\mathcal{L}^y| = \sqrt{(\mathcal{L}^y)^* \mathcal{L}^y}$$

denotes the positive square root. It is important to note that both $A$ and $\gamma$ are analytic functions of $|\mathcal{L}^y|^2 s^2$ (see Remark 8.91). We also define

(22.6)
$$k^y(s,u) = \frac{(n-1)!}{2\pi} \frac{A^y(s)}{(\gamma^y(s,u') - isu^0)^n},$$

and

(22.7)
$$K_J^y(u) = \int_{-\infty}^{\infty} k^y(s,u) \left( \prod_{j \in \{J\}} e^{-4s\lambda_j^y} \right) \left( \prod_{j \notin \{J\}} e^{4s\lambda_j^y} \right) ds.$$

Then Theorem 5.38 implies

(22.8) PROPOSITION: *Assume condition* $Y(q)$ *holds at* $y \in M$. *Set*

(22.9)
$$u_1 \cdot u_2 = (u_1^0 + u_2^0 + 2\,\mathrm{Im}\langle \mathcal{L}^y w_1, w_2 \rangle, \quad u_1' + u_2'),$$

*where* $w^j = u^j + iu^{j+n}$, $j = 1, \ldots, n$ *and* $u = (u^0, u') = (u^0, w)$. *Then*

(22.10)
$$(\square_J^y)^{-1}\phi(u) = \int_{\tilde{u}} K_J^y(\tilde{u}^{-1} \cdot u)\phi(\tilde{u})d\tilde{u}.$$

(22.11) COROLLARY: *Assume condition* $Y(q)$ *holds at* $y \in M$. *Suppose* $\square_{b,q}^y$ *is in normal form, consequently diagonal. Then*

(22.12)
$$(\square_{b,q}^y)^{-1}f(u) = \sum_{|J|=q} \left( 4 \int_{\tilde{u}} K_J^y(-\psi_u(\tilde{u}))f_J(\tilde{u})d\tilde{u} \right) d\overline{w}^J,$$

*where*

(22.13)
$$f(u) = \sum_{|J|=q} f_J(u)d\overline{w}^J.$$

(22.14) REMARK: A simple calculation yields

(22.15)
$$\tilde{u}^{-1} \cdot u = -\psi_u(\tilde{u}),$$

where, as usual, $\psi_u(\cdot)$ denotes the map to $u$-coordinates. We introduced $-\psi_u(\tilde{u})$ in preparation for the formula on a general $CR$-structure.

(22.12) has an invariant formulation. We note that

(22.16)
$$\omega = i^n \omega_u^0 \wedge dw^1 \wedge d\overline{w}^1 \wedge \cdots \wedge dw^n \wedge d\overline{w}^n$$
$$= 2^n du^0 \wedge du^1 \wedge \cdots \wedge du^{2n},$$

i.e.

(22.17)
$$\omega = 2^n du,$$

Here $U_j$, $j = 1, 2, \ldots, 2n$, is defined by (21.27) and (21.28). When condition $Y(q)$ holds $\sqcup_J^y$ is invertible for all $J$, $|J| = q$, and its inverse is given by Proposition 22.8. Again, let

$$k^y(s, u) = \frac{(n-1)!}{2\pi} \frac{A^y(s)}{(\gamma^y(s, u') - isu^0)^n},$$

and set

$$(23.1) \qquad K_J^y(u) = -\int_{-\infty}^{\infty} s \frac{\partial}{\partial s} \left\{ k^y(s, u) \det(e^{4s\mathcal{L}^y}) \left( \prod_{j \in \{J\}} e^{-8s\lambda_j^y} \right) ds \right\}.$$

When condition $Y(q)$ holds, i.e.

$$(23.2) \qquad -\sum_{j=1}^n |\lambda_j^y| < \sum_{j \in \{J\}} \lambda_j^y - \sum_{j \notin \{J\}} \lambda_j^y < \sum_{j=1}^n |\lambda_j^y|,$$

(23.1) can be integrated by parts into the form (22.7). On the other hand, (23.1) makes sense even when condition $Y(q)$ does not hold, i.e. when

$$(23.3) \qquad \sum_{j \in \{J\}} \lambda_j^y - \sum_{j \notin \{J\}} \lambda_j^y = \pm \sum_{j=1}^n |\lambda_j^y|.$$

In this case $K_J^y(\tilde{u}^{-1} \cdot u)$ is the kernel of the partial inverse of $\sqcup_J^y$ (see Theorem 7.33). To state the correct formula we need the projection onto the complement of the range of $\sqcup_J^y$. In particular we rewrite (7.25) as follows. Let

$$(23.4) \qquad j(J) = \int_{-\infty}^{\infty} \frac{\partial}{\partial s} \left\{ k^y(s, u) \det(e^{4s\mathcal{L}^y}) \left( \prod_{j \in \{J\}} e^{-8s\lambda_j^y} \right) ds \right\}.$$

We note that if condition $Y(q)$ holds, i.e. (23.2), then $j(J)$ vanishes, but if condition $Y(q)$ does not hold, i.e. (23.3), then $j(J)$ agrees with (7.25). To get the projection we differentiate under the integral sign and set

$$(23.5) \qquad \begin{aligned} S^y(s, u) &= \frac{d}{d(iu^0)} k^y(s, u) \\ &= \frac{n!}{2\pi} \frac{s A^y(s)}{(\gamma^y(s, u') - isu^0)^{n+1}}, \end{aligned}$$

and

$$(23.6) \qquad S_J^y(u) = \int_{-\infty}^{\infty} \frac{\partial}{\partial s} \left\{ S^y(s, u) \det(e^{4s\mathcal{L}^y}) \left( \prod_{j \in \{J\}} e^{-8s\lambda_j^y} \right) ds \right\}.$$

Finally we define the operators

$$(23.7) \qquad K_J^y \phi(u) = \int_{\tilde{u}} K_J^6(\tilde{u}^{-1} \cdot u) \phi(\tilde{u}) d\tilde{u},$$

where $du$ denotes Lebesgue measure. Also

$$(22.18) \qquad \left( \prod_{j \in \{J\}} e^{-4s\lambda_j^y} \right) \left( \prod_{j \notin \{J\}} e^{4s\lambda_j^y} \right) = \det([e^{-8s\mathcal{L}^y}]_{JJ}) \det(e^{4s\mathcal{L}^y}).$$

Here we used the following notation: for an arbitrary $n \times n$ matrix $A = (a_{ij})$, we set

$$(22.19) \qquad A_{IJ} = \begin{pmatrix} a_{i_1 j_1} & \cdots & a_{i_1 j_q} \\ \vdots & & \\ a_{i_q j_1} & \cdots & a_{i_q j_q} \end{pmatrix},$$

where $I = (i_1, \ldots, i_q)$ and $J = (j_1, \ldots, j_q)$. In other words $\det(A_{IJ})$ is the $(I, J)$-th element of $\Lambda^q A$. Next we define a $(2n+1)$-form $\Omega$ on $\mathbf{R}^{2n+1} \times \mathbf{R}^{2n+1}$:

$$(22.20) \qquad \Omega_{u, \tilde{u}}^y(s) = \sum_{|I| = |J| = q} \Omega_{u, \tilde{u}}^{y, IJ}(s),$$

where

$$(22.21) \qquad \begin{aligned} \Omega_{u, \tilde{u}}^{y, IJ}(s) &= 2^{-n} \epsilon(J^c, J) \\ &\quad \cdot \det(e^{4s\mathcal{L}^y}) \det([e^{-8s\mathcal{L}^y}]_{IJ}) \\ &\quad \cdot d\overline{w}^I \wedge \omega_{\tilde{u}}^0 \wedge d\tilde{w}^1 \wedge \cdots \wedge d\tilde{w}^n \wedge d\overline{\tilde{w}}^{J^c}. \end{aligned}$$

Here $J^c$ denotes the ordered complement of $J$ and $\epsilon(J^c, J)$ stands for the sign of the permutation which takes $(J^c, J)$ to $(1, 2, \ldots, n)$. Finally we set

$$(22.22) \qquad K_{u, \tilde{u}}^y(\tilde{\tilde{u}}) = \int_{-\infty}^{\infty} k^y(s, \tilde{\tilde{u}}) \Omega_{u, \tilde{u}}^y(s) ds.$$

For each $\tilde{\tilde{u}} \in \mathbf{R}^{2n+1}$, $K_{u, \tilde{u}}^y(\tilde{\tilde{u}})$ is again a $(2n+1)$-form on $\mathbf{R}^{2n+1} \times \mathbf{R}^{2n+1}$. Using this notation (22.12) can be put in the following form:

$$(22.23) \qquad (\square_{b, q}^y)^{-1} f(u) = 4 \int_{\tilde{u}} K_{u, \tilde{u}}^y(-\psi_u(\tilde{u})) \wedge f(\tilde{u}).$$

Of course, $\mathcal{L}^y$ is diagonal in the normal $u$-coordinates. On the other hand, $\Omega_{u, \tilde{u}}^y(s)$ is invariant under unitary transformations of $w$ (hence of $\tilde{w}$). In particular, changing back to skew-symmetric $v$-coordinates, i.e. $w = \overline{U} z$ (see (21.21)), we have derived

(22.24) PROPOSITION: *Let condition $Y(q)$ hold at $y \in M$. Suppose $\square_{b, q}^y$ is in skew-symmetric form, i.e. $\square_{b, q}^y$ is given by (21.10), where $X_0$ and $Z_j$, $j = 1, \ldots, n$ are in the form (21.15). Then*

$$(22.25) \qquad (\square_{b, q}^y)^{-1} f(v) = 4 \int_{\tilde{v}} K_{v, \tilde{v}}^y(-\psi_v(\tilde{v})) \wedge f(\tilde{v})$$

*with*

(22.26)
$$f(v) = \sum_{|J|=q} f_J(v) d\bar{z}^J.$$

We are ready to produce $(\Box_{b,q}^y)^{-1}$ when $\Box_{b,q}^y$ is the general model operator (21.10). All that is left is the minor complication of making the coordinate change (21.14), i.e.

(22.27)
$$\begin{cases} v^0 = x^0 - \mathrm{Re} \sum_{j,k=1}^n (a_{jk}z^j z^k + b_{jk}z^j \bar{z}^k), \\ v^j = x^j, \quad j = 1, \dots, 2n. \end{cases}$$

We shall denote this map by $v^y(x)$; of course, the $a_{jk}$'s and $b_{jk}$'s depend on $y \in M$. Then Proposition 22.24 implies

(22.28) THEOREM: *Suppose condition $Y(q)$ holds at $y \in M$. Then*

(22.29)
$$(\Box_{b,q}^y)^{-1} f(x) = 4 \int_{\tilde{x}} K_{x,\tilde{x}}^y(v^y(-\psi_x(\tilde{x}))) \wedge f(\tilde{x}),$$

*where*

$$f(x) = \sum_{|J|=q} f_J(x) d\bar{z}^J.$$

Theorem 22.28 yields the principal term in the asymptotic expansion of the kernel of $\Box_{b,q}^{-1}$ as follows. Let $U \subset M$ with local coordinates $y = (y^0, y') = (y^0, z)$. Define

(22.30)
$$\Omega_{y,\tilde{y}}(s) = \sum_{|I|=|J|=q} \Omega_{y,\tilde{y}}^{IJ}(s),$$

where

(22.31)
$$\begin{aligned} \Omega_{y,\tilde{y}}^{IJ}(s) = {}& 2^{-n} \epsilon(J^c, J) \\ & \cdot \det(e^{4s\mathcal{L}^y}) \det([e^{-8s\mathcal{L}^y}]_{IJ}) \\ & \cdot \overline{\omega}^I \wedge \omega_{\tilde{y}}^0 \wedge \omega_{\tilde{y}}^1 \wedge \cdots \wedge \omega_{\tilde{y}}^n \wedge \overline{\omega}_{\tilde{y}}^{J^c}. \end{aligned}$$

$\Omega_{y,\tilde{y}}(s)$ is a $(2n+1)$-form on $U \times U$. Following (22.22) we set

(22.32)
$$K_{y,\tilde{y}}(\tilde{\tilde{y}}) = \int_{-\infty}^{\infty} k^y(s, \tilde{\tilde{y}}) \Omega_{y,\tilde{y}}(s) ds.$$

Then Theorem 18.4 and 22.28 imply

(22.33) THEOREM: *Let condition $Y(q)$ hold at each $y \in U$. Then*

(22.34)
$$\Box_{b,q} \int_{\tilde{y}} 4 K_{y,\tilde{y}}(v^y(-\psi_y(\tilde{y}))) \wedge f(\tilde{y}) = f(y) - R_{-1} f(y),$$

*where*

(22.35)
$$f(y) = \sum_{|J|=q} f_J(y) \overline{\omega}^J, \quad f_J \in C_c^\infty(U),$$

$\psi_y(\cdot)$ *is the $y$-coordinate map,*

(22.36)
$$R_{-1} \in \mathrm{Op}\, S_{\mathcal{V}}^{-1}(U)$$

*and*

(22.37)
$$\mathcal{V} = \mathrm{Re}(T_{1,0}) + \mathrm{Im}(T_{1,0}).$$

*Furthermore, the asymptotic sum $\sum_{k=0}^{\infty} 4KR_{-1}^k$ is a parametrix for $\Box_{b,a}$, i.e.*

(22.38)
$$\Box_{b,q} \circ \sum_{k=0}^{\infty} 4KR_{-1}^k = I \mod \mathrm{Op}\, S_{\mathcal{V}}^{-\infty}(U).$$

(22.39) REMARK: We note that

(22.40)
$$K \in \mathrm{Op}\, S_{\mathcal{V}}^{-2}(U),$$

and its kernel, $K_{y,\tilde{y}}(v^y(-\psi_y(\tilde{y})))$ is invariant mod $\mathrm{Op}\, S_{\mathcal{V}}^{-3}(U)$ under diffeomorphisms of $U$.

(22.41) REMARK: We note that the integrals with respect to $s$ should be understood in the sense of (5.44). This also applies to §23.

## §23 *Partial Inverses and Projections in Case of a Non-Degenerate Levi Form*

Assuming a non-degenerate Levi form we shall translate Theorem into the terminology of $\Box_{b,q}$. We do not assume condition $Y(q)$. We by recalling the normal form of $\Box_{b,q}$ (see (22.1)):

$$\Box_{b,q}^y \left( \sum_{|J|=q} f_J d\overline{w}^J \right) = \frac{1}{4} \sum_{|J|=q} \Box_J^y(f_J) d\overline{w}^J,$$

where

$$\Box_J^y = -\sum_{j=1}^{2n} U_j^2 - i \left( \sum_{j \in \{J\}} 4\lambda_j^y - \sum_{j \notin \{J\}} 4\lambda_j^y \right) \frac{\partial}{\partial u^0}.$$

and

(23.8) $$S_J^y \phi(u) = \int_{\tilde{u}} S_J^y (\tilde{u}^{-1} \cdot u) \phi(\tilde{u}) d\tilde{u},$$

where $u_1 \cdot u_2$ is given by (22.9). Then Theorem 7.33 translates into

(23.9) THEOREM: *Assume that the Levi form, $\mathcal{L}^y$, is non-degenerate at $y \in M$. Then*

(23.10) $$\Box_J^y K_J^y + S_J^y = I, \quad |J| = q.$$

*If*

(23.11) $$\left| \sum_{j \in \{J\}} \lambda_j^y - \sum_{j \notin \{J\}} \lambda_j^y \right| < \sum_{j=1}^n |\lambda_j^y|,$$

*then $S_J^y$ vanishes and (23.10) agrees with (22.10).*

(23.12) REMARK: The importance of Theorem 23.9 stems from the fact that (23.10) holds even at the end-points of the basic interval,

$$\left[ -\sum_{j=1}^n |\lambda_j^y|, \ \sum_{j=1}^n |\lambda_j^y| \right],$$

if $K_J^y$ and $S_J^y$ are defined by (23.1) and (23.6), respectively.

(23.13) COROLLARY: *Assume a non-degenerate Levi form $\mathcal{L}^y$ at $y \in M$. Suppose $\Box_{b,q}^y$ is in normal form (see (22.1)). Then*

(23.14)
$$\Box_{b,q}^y \left( \sum_{|J|=q} \left( 4 \int_{\tilde{u}} K_J^y (-\psi_u(\tilde{u})) f_J(\tilde{u}) d\tilde{u} \right) d\overline{w}^J \right)$$
$$+ \sum_{|J|=q} \left( \int_{\tilde{u}} S_J^y (-\psi_u(\tilde{u})) f_J(\tilde{u}) d\tilde{u} \right) d\overline{w}^J$$
$$= \sum_{|J|=q} f_J(u) d\overline{w}^J,$$

*where $K_J^y$ and $S_J^y$ are given by (23.1) and (23.6), respectively. If condition $Y(q)$ holds at $y \in M$, then the second sum on the left-hand side of (23.14) vanishes and we are reduced to Corollary 22.11.*

Following the discussion of §22 we can easily transfer this result to a $CR$-structure. Again, we set

$$\Omega_{y,\tilde{y}}(s) = \sum_{|I|=|J|=q} \Omega_{y,\tilde{y}}^{IJ}(s),$$

where $\Omega_{y,\tilde{y}}^{IJ}(s)$ is defined by (22.31). Also set

$$(23.15) \qquad K_{y,\tilde{y}}(\tilde{\tilde{y}}) = \int_{-\infty}^{\infty} s \frac{\partial}{\partial s} \{ k^y(s, \tilde{\tilde{y}}) \Omega_{y,\tilde{y}}(s) \} ds,$$

where $k^y(s, \tilde{\tilde{y}})$ is given by (22.6). Similarly, we have

$$(23.16) \qquad S_{y,\tilde{y}}(\tilde{\tilde{y}}) = \int_{-\infty}^{\infty} \frac{\partial}{\partial s} \{ S^y(s, \tilde{\tilde{y}}) \Omega_{y,\tilde{y}}(s) \} ds,$$

where $S^y(s, \tilde{\tilde{y}})$ is defined by (23.5). Again, we let $v^y(x)$ denote the map (22.27). Then we have derived

(23.17) THEOREM: *Assume the Levi form $\mathcal{L}^y$ is non-degenerate at all points $y \in M$. Then*

$$(23.18) \qquad \begin{aligned} \Box_{b,q} \int_{\tilde{y}} 4 K_{y,\tilde{y}}(v^y(-\psi_y(\tilde{y}))) \wedge f(\tilde{y}) \\ + \int_{\tilde{y}} S_{y,\tilde{y}}(v^y(-\psi_y(\tilde{y}))) \wedge f(\tilde{y}) \\ = f(y) - R_{-1}f(y), \end{aligned}$$

*where*

$$(23.19) \qquad f(y) = \sum_{|J|=q} f_J(y) \overline{w}^J \in C_c^\infty(M; \Lambda^{0,q}),$$

$$(23.20) \qquad R_{-1} \in \mathrm{Op}\, S_{\mathcal{V}}^{-1}(M), \quad \mathcal{V} = \mathrm{Re}(T_{1,0}) + \mathrm{Im}(T_{1,0}),$$

and $K_{y,\tilde{y}}(\tilde{\tilde{y}})$, $S_{y,\tilde{y}}(\tilde{\tilde{y}})$ are defined by (23.15) and (23.16), respectively. If condition $Y(q)$ holds at each point $y \in M$, then $S_{y,\tilde{y}}(\tilde{\tilde{y}})$ vanishes and (23.18) becomes (21.34).

(23.21) REMARK: $4K_{y,\tilde{y}}(v^y(-\psi_y(\tilde{y})))$ and $S_{y,\tilde{y}}(v^y(-\psi_y(\tilde{y})))$ represent the principal term of a *"partial inverse"* and of a *"projection"* acting on $L^2(M; \Lambda^{0,q})$. This will be treated more fully in §24 and §25.

## §24 $\Box_b$ on a Compact Manifold

Suppose that $M$ is a compact $CR$-manifold of dimension $2n+1$ with hermitian metric. Let $\overline{\partial}_{b,q}$ be the corresponding operator on $(0,q)$-forms and $\vartheta_{b,q+1}$ the formal adjoint acting on $(0, q + 1)$-forms. Again

(24.1)
$$\square_{b,q} = \vartheta_{b,q+1}\overline{\partial}_{b,q} + \overline{\partial}_{b,q-1}\vartheta_{b,q}$$
$$= \square_{b,q}^*.$$

(24.2) THEOREM: *Suppose $0 \le q \le n$ and condition $Y(q)$ is satisfied at each point of $M$. Then the $L^2$-realization $[\square_{b,q}]$ has the following properties. It is self-adjoint in $L^2(M; \Lambda^{0,q})$, has finite-dimensional kernel, and has closed range with finite codimension. The partial inverse $N_{b,q}$ is a $\mathcal{V}$-operator of order $-2$.*

Proof. According to Theorem 22.33 $Y(q)$ implies the existence of local parametrices which are $\mathcal{V}$-operators of order $-2$. By Remark 19.14 there is a global parametrix $Q$ which is a $\mathcal{V}$-operator of order $-2$. The remaining assertions follow from Theorem 19.16. The proof of Theorem 19.16 shows that $N_{b,q}$ differs from $Q$ by a smoothing operator.

We may also obtain information about separate pieces of $\square_{b,q}$, using Theorem 19.8 and some algebra. We begin with the algebra. Set

(24.3)          $$\square_{b,q}' = \vartheta_{b,q+1}\overline{\partial}_{b,q}, \quad \square_{b,q}'' = \overline{\partial}_{b,q-1}\vartheta_{b,q}.$$

Recall that

(24.4)          $$\overline{\partial}_{b,q+1}\overline{\partial}_{b,q} = 0, \quad \text{all} \quad q,$$

so taking the formal adjoint

(24.5)          $$\vartheta_{b,q-1}\vartheta_{b,q} = 0, \quad \text{all} \quad q.$$

Therefore

(24.6)
$$\overline{\partial}_{b,q}\square_{b,q} = \overline{\partial}_{b,q}\square_{b,q}' = \square_{b,q+1}''\overline{\partial}_{b,q}$$
$$= \square_{b,q+1}\overline{\partial}_{b,q}, \quad \text{all} \quad q,$$

(24.7)
$$\vartheta_{b,q}\square_{b,q} = \vartheta_{b,q}\square_{b,q}'' = \square_{b,q-1}'\vartheta_{b,q}$$
$$= \square_{b,q-1}\vartheta_{b,q}, \quad \text{all} \quad q,$$

(24.8)          $$\square_{b,q}\square_{b,q}' = \square_{b,q}'\square_{b,q}, \quad \square_{b,q}\square_{b,q}'' = \square_{b,q}''\square_{b,q}.$$

We shall need some analogous identities involving the partial inverse $N_{b,q}$ and one of the associated projections.

(24.9) LEMMA: *Suppose condition $Y(q)$ is satisfied at every point of $M$. Let $N = N_{b,q}$ be the partial inverse of $\square_{b,q}$ and let $\Pi_1$ be the orthogonal* *projection on* $\mathrm{ran}[\square_{b,q}] = (\mathrm{ker}[\square_{b,q}])^{\perp}$. *Then as* $\mathcal{V}$-*operators:*

(24.10)
$$\overline{\partial}_{b,q}\Pi_1 = \overline{\partial}_{b,q},$$

(24.11)
$$\vartheta_{b,q}\Pi_1 = \vartheta_{b,q},$$

(24.12)
$$\Pi_1\vartheta_{b,q+1} = \vartheta_{b,q+1},$$

(24.13)
$$\Pi_1\overline{\partial}_{b,q-1} = \overline{\partial}_{b,q-1},$$

(24.14)
$$\Pi_1\square'_{b,q} = \square'_{b,q} = \square'_{b,q}\Pi_1,$$

(24.15)
$$\Pi_1\square''_{b,q} = \square''_{b,q} = \square''_{b,q}\Pi_1,$$

(24.16)
$$N\square'_{b,q} = \square'_{b,q}N,$$

(24.17)
$$N\square''_{b,q} = \square''_{b,q}N.$$

Proof. For $u \in \mathrm{ker}[\square_{b,q}] \subset C^{\infty}(M'\Lambda^{0,q})$,

(24.18)
$$0 = (\square_{b,q}u, u) = (\overline{\partial}_{b,q}u, \overline{\partial}_{b,q}u) + (\vartheta_{b,q}u, \vartheta_{b,q}u).$$

Therefore

$$\mathrm{ker}[\square_{b,q}] = \mathrm{ker}(\overline{\partial}_{b,q}) \cap \mathrm{ker}(\vartheta_{b,q}) \cap C^{\infty}(M; \Lambda^{0,q}).$$

Now $\Pi_1$ is a $\mathcal{V}$-operator with

$$\mathrm{ker}[\Pi_1] = \mathrm{ker}[\square_{b,q}].$$

Therefore (24.10) and (24.11) hold on $C^{\infty}(M; \Lambda^{0,q})$, so they hold identically. Taking the formal adjoints, we obtain (24.12) and (24.13).The identities (24.14), (24.15) follow from (24.10)–(24.13). Finally,

$$\begin{aligned} N\square'_{b,q} &= N\square'_{b,q}\Pi_1 = N\square'_{b,q}\square_{b,q}N \\ &= N\square_{b,q}\square'_{b,q}N = \Pi_1\square'_{b,q}N \\ &= \square'_{b,q}N \end{aligned}$$

and similarly for (24.17).

(24.19) REMARK: In writing operators like $\overline{\partial}_{b,q}N$ or $N\overline{\partial}_{b,q-1}$ in what follows, we may omit the subscripts without ambiguity, since $N$ acts only in $\mathcal{D}'(M; \Lambda^{0,q})$.

(24.20) THEOREM: *Suppose* $0 \leq q \leq n$ *and condition* $Y(q)$ *is satisfied at each point of* $M$. *Consider the* $L^2$ *realization of each operator in the* *first column of the table below. Then each has closed range, and the partial inverse and associated projections are the $L^2$-realizations of the indicated $\mathcal{V}$-operators. Moreover, (ii) is the adjoint of (i), (iv) is the adjoint of (iii), and (v)–(viii) are self-adjoint. In each case $N = N_{b,q}$ denotes the partial inverse of $\square_{b,q}$, and for convenience we omit subscripts.*

|  | Operator | Partial Inverse | Projection to Range | Projection to (ker)$^{\perp}$ |
|---|---|---|---|---|
| (i) | $\overline{\partial}_{b,q}$ | $N\vartheta$ | $\overline{\partial}N\vartheta$ | $N\square'$ |
| (ii) | $\vartheta_{b,q+1}$ | $\overline{\partial}N$ | $\square'N$ | $\overline{\partial}N\vartheta$ |
| (iii) | $\overline{\partial}_{b,q-1}$ | $\vartheta N$ | $\square''N$ | $\vartheta N\overline{\partial}$ |
| (iv) | $\vartheta_{b,q}$ | $N\overline{\partial}$ | $\vartheta N\overline{\partial}$ | $N\square''$ |
| (v) | $\square'_{b,q}$ | $N\square'N$ | $\square'N$ | $N\square'$ |
| (vi) | $\square''_{b,q}$ | $N\square''N$ | $\square''N$ | $N\square''$ |
| (vii) | $\square'_{b,q-1}$ | $\vartheta N^2\overline{\partial}$ | $\vartheta N\overline{\partial}$ | $\vartheta N\overline{\partial}$ |
| (viii) | $\square''_{b,q+1}$ | $\overline{\partial}N^2\vartheta$ | $\overline{\partial}N\vartheta$ | $\overline{\partial}N\vartheta$ |

*(we take the operators in (iii) and (vii) to be zero if $q = 0$, and in (ii) and (viii) to be zero if $q = n$).*

Proof. As noted above, we use Theorem 19.8 and the accumulated identities above. Our first step is to establish the identities (19.9)–(19.11) in the present context. Each of our proposed projections is formally self-adjoint, because $\vartheta = \overline{\partial}^*$, $N = N^*$, $\square' = (\square')^*$, $\square'' = (\square'')^*$, and $N$ commutes with $\square'$ and $\square''$. To show that these proposed projections are idempotents we treat two representative cases. First,

$$(\overline{\partial}N\vartheta)^2 = \overline{\partial}N\square'N\vartheta = \overline{\partial}N^2\square'\vartheta = \overline{\partial}N^2\square\vartheta = \overline{\partial}N\Pi_1\vartheta = \overline{\partial}N\vartheta.$$

Secondly

$$(N\square')^2 = N\square'N\square' = N^2\square'\square' = N^2\square\square' = N\Pi_1\square' = N\square'.$$

In cases (i)–(iv) the identities (19.9) are immediate, and in cases (v) and (vi) they follow from the fact that $\square'N = N\square'$ and $\square''N = N\square''$ are idempotent. For case (vii) we have

$$\square'(\vartheta N^2\overline{\partial}) = \vartheta\square'N^2\overline{\partial} = \vartheta\square N^2\overline{\partial} = \vartheta\Pi_1N\overline{\partial} = \vartheta N\overline{\partial};$$
$$(\vartheta N^2\overline{\partial})\square' = \vartheta N^2\square'\overline{\partial} = \vartheta N^2\square\overline{\partial} = \vartheta N\Pi_1\overline{\partial} = \vartheta N\overline{\partial}.$$

The argument in case (viii) is similar.

Finally, we must establish the identities (19.10). We consider three representative cases. In case (i),

$$(\overline{\partial}N\vartheta)\overline{\partial} = \overline{\partial}N\square' = \overline{\partial}\square'N = \overline{\partial}\square N = \overline{\partial}\Pi_1 = \overline{\partial}.$$

In case (v),

$$(\Box'N)\Box' = \Box'(N\Box') = (\Box')^2 N = \Box'\Box N = \Box'\Pi_1 = \Box'.$$

In case (vii) we need two computations:

$$\Box'(\vartheta N\overline{\partial}) = \vartheta\Box'N\overline{\partial} = \vartheta\Box N\overline{\partial} = \vartheta\Pi_1\overline{\partial} = \vartheta\overline{\partial} = \Box';$$
$$(\vartheta N\overline{\partial})\Box' = \vartheta N\Box'\overline{\partial} = \vartheta N\Box\overline{\partial} = \vartheta\Pi_1\overline{\partial} = \vartheta\overline{\partial} = \Box'.$$

The other cases are handled similarly.

(24.21) THEOREM: *Suppose $0 < q < n$ and conditions $Y(q-1)$ and $Y(q+1)$ are satisfied at every point of $M$. Then the $L^2$-realization $[\Box_{b,q}]$ is self-adjoint and has closed range. The partial inverse is*

(24.22)             $$\vartheta_{b,q+1}N^2_{b,q+1}\overline{\partial}_{b,q} + \partial_{b,q-1}N^2_{b,q-1}\vartheta_{b,q},$$

*and the orthogonal projection onto the range is*

(24.23)             $$\vartheta_{b,q+1}N_{b,q+1}\overline{\partial}_{b,q} + \overline{\partial}_{b,q-1}N_{b,q-1}\vartheta_{b,q},$$

*where $N_{b,q-1}$ and $N_{b,q+1}$ are the partial inverses of $\Box_{b,q-1}$ and $\Box_{b,q+1}$, respectively.*

Proof. Let $P' = [\Box'_{b,q}]$, $P'' = [\Box''_{b,q}]$, $A' = [\vartheta_{b,q+1}N^2_{b,q+1}\overline{\partial}_{b,q}]$, $A'' = [\overline{\partial}_{b,q-1}N^2_{b,q-1}\vartheta_{b,q}]$ and $\Pi' = [\vartheta_{b,q+1}N_{b,q+1}\overline{\partial}_{b,q}]$, $\Pi'' = [\overline{\partial}_{b,q-1}N_{b,q-1}\vartheta_{b,q}]$. Note that $\Box'_{b,q}$ is operator (vii) in the case $q + 1$, so $P'$ is self-adjoint and has closed range, while $A'$ is the partial inverse and $\Pi'$ is the associated projection. Similarly, $\Box''_{b,q}$ is operator (viii) for the case $q - 1$, so $P''$ is self-adjoint and has closed range, while $A''$ is the partial inverse and $\Pi''$ the associated projection. Since $P'P'' = P''P' = 0$, and $P', P''$ are self-adjoint,

$$\text{ran } P' = (\ker P')^{\perp} \subset (\text{ran } P'')^{\perp}.$$

Therefore $\text{ran } P' + \text{ran } P'' = \text{ran}(P' + P'')$ is closed and it is easy to check that $A' + A''$ is the partial inverse and $\Pi' + \Pi''$ is the associated projection. The desired conclusions now follow from Theorem 19.8.

(24.24) REMARKS: We have now shown that the $L^2$-realization of $\Box_{b,q}$ has a $\mathcal{V}$-operator as partial inverse, $N_{b,q}$, in each of the following cases:

(24.25)     *$Y(q)$ holds at each point of $M$; then $N_{b,q}$ is a parametrix.*

(24.26)     *$0 < q < n$ and both $Y(q-1)$ and $Y(q+1)$ hold at each point of $M$.*

(24.27)     *$q = 0$ and $Y(1)$ holds at each point of $M$: this is case (vii) of Theorem 24.20.*

(24.28)    $q = n$ and $Y(n-1)$ holds at each point of $M$: this is case (viii) of Theorem 24.20.

(24.29) PROPOSITION: *Suppose $M$ is connected and one of (24.26), (24.27) or (24.28) is true. Then either $Y(q)$ holds at each point of $M$ or at each point of $M$ the Levi form is non-degenerate and $Y(q)$ fails.*

Proof. Let $\{\lambda_1, \ldots, \lambda_n\}$ be the eigenvalues of the Levi form at a point $x \in M$. Suppose (24.26) is satisfied. If $Y(q)$ fails at $x$ then after renumbering we may assume that $-\lambda_1, -\lambda_2, \ldots, -\lambda_q, \lambda_{q+1}, \ldots, \lambda_n$ are all $\geq 0$ or are all $\leq 0$. If one of $\lambda_1, \ldots, \lambda_q$ were zero, then $Y(q-1)$ would fail. If one of $\lambda_{q+1}, \ldots, \lambda_n$ were zero, then $Y(q+1)$ would fail. Therefore the set of $x$ where $Y(q)$ fails is open and the Levi form is nondegenerate at each such point. The set where $Y(q)$ fails is easily seen to be closed as well. This completes the proof in case (24.26). The argument is essentially the same in cases (24.27) and (24.28).

When (24.25) fails but one of (24.26)–(24.28) holds, we can, in principle, determine a complete asymptotic expansion for $N_{b,q}$ from the expansions of the parametrices for $\Box_{b,q-1}$ (if $q > 0$) and $\Box_{b,q+1}$ (if $q < n$), since these give $N_{b,q-1}$ and $N_{b,q+1}$. Note that the constructions in §23 give candidates for the principal terms of $N_{b,q}$ and the associated projection $\mathrm{II}_1$ (or $I - \mathrm{II}_1$). In the next section we give another asymptotic construction, which works in greater generality and shows that our candidates for the principal terms are indeed the principal symbols.

## §25 *The Partial Inverse and Associated Projections for $\Box_{b,q}$*

In this section we give a different approach to the asymptotic construction of the partial inverse $N_{b,q}$ and associated projections. We begin by reducing the construction of a formal partial inverse and formal projections to the following problem: "*let us look for $\mathcal{V}$-operators $\tilde{S}$ and $\tilde{Q}$ such that*

(25.1)          $\tilde{S}$ *has order* $0$, $\tilde{Q}$ *has order* $-2$,

(25.2)          $\Box_{b,q}\tilde{Q} + \tilde{S} = I - R,$

*where $R$ has order* $-1$, *and*

(25.3)          $\Box_{b,q}\tilde{S} \sim 0$."

(As in Chapter 3 we write $A \sim B$ if $A - B$ is a smoothing operator.)

(25.4) PROPOSITION: *Suppose $\tilde{S}$ and $\tilde{Q}$ are $\mathcal{V}$-operators which satisfy (25.1)–(25.3). Then there are $\mathcal{V}$-operators $S$ and $Q$ such that $S - \tilde{S}$ has order $-1$, $Q - \tilde{Q}$ has order $-3$ and*

$$(25.5) \qquad \qquad \Box_{b,q} Q + S \sim I,$$

$$(25.6) \qquad \qquad S \sim S^* \sim S^2, \quad \Box_{b,q} S \sim S \Box_{b,q} \sim 0,$$

$$(25.7) \qquad \qquad SQ \sim QS \sim 0, \quad and$$

$$(25.8) \qquad \qquad Q \sim Q^*, \quad \Box_{b,q} Q \sim Q \Box_{b,a}.$$

*Moreover, (25.5)–(25.8) determine $S$ and $Q$ up to smoothing operators.*

Proof. Let $A$ be a parametrix for $I - R$ and set

$$(25.9) \qquad \qquad S = \tilde{S} A, \quad Q = (I - S) \tilde{Q} A.$$

To simplify notation we set $\Box = \Box_{b,q}$. Then (25.2) and (25.3) imply

$$(25.10) \qquad \qquad \Box S \sim 0, \quad S^* \Box \sim 0,$$

$$(25.11) \qquad \qquad \Box Q + S \sim \Box \tilde{Q} A + \tilde{S} A \sim I,$$

hence

$$(25.12) \qquad \qquad Q^* \Box + S^* \sim I.$$

Composing (25.11) on the left by $S^*$ and (25.12) on the right by $S$ gives

$$(25.13) \qquad \qquad S^* \sim S^* S \sim S,$$

which proves (25.5) and (25.6). Now

$$(25.14) \qquad \qquad SQ = (S - S^2) \tilde{Q} A \sim 0,$$

which implies

$$(25.15) \qquad QS \sim (Q^* \Box + S) QS \sim Q^* \Box QS \sim Q^*(I - S)S \sim 0.$$

Hence we have derived (25.7). Now (25.8) follows from

$$(25.16) \qquad Q \sim (Q^* \Box + S) Q \sim Q^* \Box Q = (Q^* \Box Q)^* \sim Q^*,$$

and

$$(25.17) \qquad \qquad \Box Q \sim I - S \sim I - S^* = Q^* \Box \sim Q \Box.$$

Finally, suppose $S'$ and $Q'$ also satisfy (25.5)–(25.8). Then

(25.18)          $S' \sim (Q\square + S)S' \sim SS' \sim S(\square Q' + S) \sim S,$

(25.19)   $Q' \sim (Q\square + S)Q' \sim Q(I - S') + SQ' \sim Q(I - S) + S'Q' \sim Q.$

(25.20) THEOREM: *Suppose the $L^2$-realization of $\square_{b,q}$ has closed range. Let $N$ be the partial inverse and let $\Pi_0$ denote the orthogonal projection onto $\ker \square_{b,q}$. Suppose that $Q$ and $S$ are $\mathcal{V}$-operators of order $-2$ and $0$, respectively, which satisfy (25.5)–(25.8). Then $N$ and $\Pi_0$ are $\mathcal{V}$-operators and $N \sim Q$, $\Pi_0 \sim S$.*

Proof. For convenience we may replace $S$ by $I - \square Q$, which differs from $S$ by a smoothing operator, and have

(25.21)          $\square Q + S = I = Q^*\square + S^*.$

In what follows, the $R_j$ denote smoothing operators. We have, on the space $C^\infty(M; \Lambda^{0,q})$,

(25.22)          $S = (N\square + \Pi_0)S = \Pi_0 S + NR_1,$

(25.23)          $\Pi_0 = \Pi_0(\square Q + S) = \Pi_0 S.$

Therefore

(25.24)          $S - \Pi_0 = S - \Pi_0 S = NR_1,$

(25.25)          $S^*\Pi_0 = \Pi_0^* = \Pi_0 = \Pi_0^2,$

and

(25.26)
$$S - \Pi_0 = S^*S - S^*\Pi_0 - \Pi_0 S + \Pi_0^2 + S - S^*S$$
$$= (S^* - \Pi_0)(S - \Pi_0) + (S - S^*S)$$
$$= R_1^* N^2 R_1 + R_2.$$

Now

(25.27)          $R_1^* N^2 R_1 : \mathcal{D}'(M; \Lambda^{0,q}) \to L^2(M; \Lambda^{0,q}) \to C^\infty(M; \Lambda^{0,q}),$

so $S - \Pi_0$ is smoothing and $\Pi_0$ is a $\mathcal{V}$-operator. Next

(25.28)
$$N - Q = N(\square Q + S) - Q = NS - \Pi_0 Q$$
$$= N(S - \Pi_0) + (S - \Pi_0)Q - SQ$$
$$= NR_3 + R_4.$$

But $N = N^*$, so we also have

$$(25.29) \quad \begin{aligned} N - Q^* &= R_3^* N + R_4^* = R_3^* (Q + N R_3 + R_4) + R_4^* \\ &= R_3^* N R_3 + R_5. \end{aligned}$$

Thus $N - Q^*$ is smoothing, so $N$ is a $\mathcal{V}$-operator and $N \sim Q^* \sim Q$.

These results suggest an alternative procedure for getting the asymptotic expansion of the partial inverse and associated projections. It applies to all the cases handled in §24, i.e. cases (24.25)–(24.28), and to the classical case of $n - 1$ and $q = 0$, which does not fit the framework of §24. To see this, we note first that when condition $Y(q)$ holds we may take $\tilde{S} = 0$ and let $\tilde{Q}$ be a parametrix in (25.2). Next, suppose that $Y(q)$ fails at a point, $M$ is connected and one of (24.26)–(24.28) holds. Then, according to Proposition 24.29 the Levi form is non-degenerate at each point of $M$. In this case the construction of §23 gives $\mathcal{V}$-operators $S_0$ of order 0 and $\tilde{Q}$ of order -2 so that

$$(25.30) \quad \quad \Box \tilde{Q} + S_0 - I \quad \text{has order} \quad -1,$$

$$(25.31) \quad \quad \Box S_0 \quad \text{has order} \quad 1.$$

If we can find $\tilde{S}$ such that $\Box \tilde{S} \sim 0$ and $\tilde{S} - S_0$ has order -1, then the preceding results of this section apply. We shall show below how to construct such an $\tilde{S}$ in the classical case, $n = 1$ and $q = 0$, which is not covered by §24. The method clearly carries over to arbitrary $n$ and $q$ when the Levi form is non-degenerate.

When the Levi form is positive, condition $Y(0)$ fails. Condition $Y(1)$ holds if $n > 1$, but fails if $n = 1$. An example of Nirenberg [2] shows that for an abstract $CR$ manifold with $n = 1$ and positive Levi, the range of $\Box_b$ is not necessarily closed. However if $M$ is the boundary of a strictly pseudo-convex domain in $\mathbf{C}^2$, $\Box_b$ does have closed range but the results of §24 do not apply. The following argument shows that the results of this section do apply, so that even in this case the Cauchy–Szegó projection is a $\mathcal{V}$-operator (we refer to Theorem 25.20).

Let $n = 1$. Set

$$(25.32) \quad \quad Z_\lambda = \frac{\partial}{\partial z} - \frac{1}{2} i \lambda w \frac{\partial}{\partial x^0},$$

and let $H^{(\lambda)}$ denote the Heisenberg group which has $Z_\lambda, \overline{Z}_\lambda$ and $\partial/\partial u^0$ for a basis for its left-invariant vector fields. If we change variables: $(\lambda/2)^{\frac{1}{2}} w \to w$, i.e. $H_1^{(\lambda)} \to H_1$, a simple consequence of (13) in Greiner-Kohn-Stein [1] is

(25.33) THEOREM: *Let*

(25.34)
$$W_\lambda = \frac{\frac{1}{2}\lambda w}{\pi^2 \left(\frac{1}{4}\lambda^2 |w|^4 + (u^0)^2\right)^2}$$

*induce a left-invariant convolution operator on $H_1^{(\lambda)}$. Then*

(25.35)
$$Z_\lambda W_\lambda = I - S_{\lambda;-},$$

(25.36)
$$W_\lambda Z_\lambda = I - S_{\lambda;+},$$

*where $S_{\lambda;\pm}$ denote the Cauchy–Szegő kernels:*

(25.37)
$$S_{\lambda;\pm} = \frac{\frac{1}{2}\lambda}{\pi^2 \left(\frac{1}{2}\lambda |w|^2 \mp iu^0\right)^2}.$$

When $n = 1$, $\Box_b = \vartheta_b \overline{\partial}_b$. Consequently the projection $\Pi_0$ onto $\ker \Box_b$ is the projection onto $\ker \overline{\partial}_b$. Hence we shall construct $S \in S_{\mathcal{V}}^0(M)$ such that

(25.38)
$$\overline{\partial}_b S \sim 0,$$

Actually, we shall consruct $S^* \in S_{\mathcal{V}}^0(M)$, such that

(25.39)
$$S^* \vartheta_b \sim 0,$$

which is equivalent to (25.38). Let us suppose that $S^*$ is induced by the kernel

(25.40)
$$S^*(y, v^y(-\psi_y(\tilde{y}))),$$

where $y$ denotes local coordinates, $\psi_y(\cdot)$ is the map to $y$-coordinates and $v^y$ is given by (22.27). Set

(25.41)
$$t = \psi_y(\tilde{y}).$$

Evaluating $S^* \vartheta_b$ we may fix the first variable of $S^*$ at $y$, i.e. $t = 0$ (see Theorem 14.7 on the composition of $\mathcal{V}$-operators). Next we note that the above quadratic coordinate change, $v^y(\cdot)$, reduces the principal term of $\vartheta_b$ to skew-symmetric form (in this case, $n = 1$, to normal form). Again, we let $u$ denote the system of normal coordinates. Then $\vartheta_b$ has the following asymptotic representation in a neighborhood of $u = 0$:

(25.42) $\quad \vartheta_b \sim -Z_\lambda + \displaystyle\sum_{m=1}^{\infty} \left[ h_m^{(Z_\lambda)}(u)Z_\lambda + h_m^{(\overline{Z}_\lambda)}(u)\overline{Z}_\lambda + h_{m+1}^{(U_0)}(u)\frac{\partial}{\partial u^0} \right],$

where $h_m^{(\cdot)}$ denotes an $H$-homogeneous polynomial of degree $m$ in $u$, $\lambda = \lambda^y = \mathcal{L}^y > 0$ is the Levi form at $y$, $w = u_1 + iu_2$ and $U_0 = \frac{\partial}{\partial u^0}$. More concisely we write

$$(25.43) \qquad \vartheta_b \sim \vartheta_{b;1} + \sum_{k=0}^{\infty} \vartheta_{b;-k},$$

where

$$(25.44) \qquad \vartheta_{b;-k} = h_{k+1}^{(Z_\lambda)}(u)Z_\lambda + h_{k+1}^{(\overline{Z}_\lambda)}(u)\overline{Z}_\lambda + h_{k+2}^{(U_0)}(u)\frac{\partial}{\partial u^0}.$$

Similarly,

$$(25.45) \qquad S^*(u) = S^*(y, u) \sim S_0^*(u) + \sum_{k=1}^{\infty} S_{-k}^*(u), \quad \hat{S}_j^* \in \mathcal{G}_j,$$

where $\mathcal{G}_j$, $j \in Z$, is given by Definition 15.6. We set $S_0^* = S_{0;+}^* + S_{0;-}^*$, where

$$(25.46) \qquad S_{0;+}^* = 0,$$

$$(25.47) \qquad S_{0;-}^* = S_{\lambda;-},$$

and note that by Theorem 7.33

$$(25.48) \qquad S_{\lambda;-}(u)Z_\lambda = 0.$$

We write (25.48) as

$$(25.49) \qquad S_0^* \vartheta_{b;1} = 0.$$

To find $S_{-k}^*$, $k = 1, 2, \ldots$, we use Theorem 14.7, i.e.

$$(25.50) \qquad \left( S_0^* + \sum_{k=1}^{\infty} S_{-k}^* \right) \circ \left( \vartheta_{b;1} + \sum_{k=0}^{\infty} \vartheta_{b;-k} \right) \sim 0.$$

For example, $S_{-1}^*$ must satisfy

$$(25.51) \qquad S_{-1}^* \vartheta_{b;1} + S_0^* \vartheta_{b;0} = 0,$$

or

$$(25.52) \qquad S_{-1}^* Z_\lambda = S_0^* \vartheta_{b;0}.$$

According to Theorem 25.33

$$W_\lambda Z_\lambda = I - S_{\lambda;+}.$$

Therefore, setting

(25.53) $$S^*_{-1} = S^*_0 \vartheta_{b;0} W_\lambda,$$

we have

(25.54) $$S^*_{-1}\vartheta_{b;1} + S^*_0\vartheta_{b;0} = -S^*_0\vartheta_{b;0}(I - S_{\lambda;+}) + S^*_0\vartheta_{b;0}$$
$$= 0,$$

as required, since

(25.55) $$S^*_0\vartheta_{b;0}S_{\lambda;+} = 0.$$

Indeed, according to (2.27) of Beals–Greiner–Vauthier [1]:

(25.56) $$S^*_{0;+} = 0 \Longrightarrow (S^*_0\vartheta_{b;0})_+ = S^*_{0;+}\vartheta_{b;0;+} = 0$$
$$\Longrightarrow S^*_0\vartheta_{b;0}S_{\lambda;+} = (S^*_0\vartheta_{b;0})_+ S_{\lambda;+} = 0.$$

In general, we require

(25.57) $$\sum_{j=0}^{k} S^*_{-j}\vartheta_{b;j-k} - S^*_{-k-1}Z_\lambda = 0.$$

Therefore, the same argument yields

(25.58) $$S^*_{-k-1} = \sum_{j=0}^{k} S^*_{-j}\vartheta_{b;j-k} W_\lambda.$$

Now $\lambda = \lambda^y = \mathcal{L}^y$, which varies smoothly with $y \in M$. Thus we have derived

(25.59) THEOREM: *Let $n = 1$. There is a $\mathcal{V}$-operator $S^* \in \mathrm{Op}\, S^0_\mathcal{V}(M)$, $\mathcal{V} = \mathrm{Re}(T_{1,0}) + \mathrm{Im}(T_{1,0})$, with kernel $S^*(y, v^y(-\psi_y(\tilde{y})))$, such that*

(25.60) $$S^*\vartheta_b \sim 0.$$

*Moreover, $S^*$ has the following asymptotic expansion:*

(25.61) $$S^* \sim S^*_0 + \sum_{k=1}^{\infty} S^*_{-k}, \quad S^*_{-k} \in \mathrm{Op}\, S_{\mathcal{V}, -k}(M),$$

$k = 0, 1, 2, \ldots$, *where*

(25.62) $$S^*_0(y, u) = \frac{\frac{1}{2}\lambda^y}{\pi^2 \left(\frac{1}{2}\lambda^y|w|^2 + iu^0\right)^2}$$

*with $w = u_1 + iu_2$, $v^y(\cdot)$ the map (22.27), $\lambda^y = \mathcal{L}^y$ and $S^*_{-k}$, $k = 1, 2, \ldots$ given by (25.58).*

Since

(25.63) $$S_0 = S_0^* \quad \mod S_{\mathcal{V}}^{-1}(M),$$

we also have

(25.64) COROLLARY: *Let $n = 1$. Let $S$ denote the adjoint of $S^*$. Then*

(25.65) $$\overline{\partial}_b S \sim 0,$$

*where*

(25.66) $$S = S_0^* \quad \mod S_{\mathcal{V}}^{-1}(M).$$

(25.67) COROLLARY: *Let $n = 1$, $\mathcal{L}^y > 0$. Let $\tilde{S}$ denote $S$ of (25.66) and let $\tilde{Q}$ be induced by $K_{y,\tilde{y}}(\tilde{\tilde{y}})$ of (23.15). Then (25.1)–(25.3) are satisfied. Consequently Proposition 25.4 and Theorem 25.20 hold. In particular the Cauchy–Szegő projection $C = \Pi_0$ is a $\mathcal{V}$-operator and $C \sim \tilde{S}A$.*

(25.68) REMARK: As we pointed out earlier, this construction of $\tilde{S}$ also works for $\square_{b,q}$ in all the non-hypoelliptic cases of §24, i.e. in the cases (24.26)–(24.28).

(25.69) REMARK: If $S^*$ is given by (25.61) then the actual kernel is $|\psi_y'|^{-1} S^*(y, v^y(-\psi_y(\tilde{y})))$.

# Bibliography

M. Beals, C. Fefferman and R. Grossman

[1]. *Strictly pseudoconvex domains in* $C^n$, Bull. Amer. Math. Soc. 8 (1983), 125–322.

R. Beals

[1]. *A characterization of pseudodifferential operators and applications*, Duke Math. J. 44 (1977), 45–57; Correction, *ibid* 46 (1979), 215.

[2]. $L^p$ *and Hölder estimates for pseudodifferential operators: sufficient conditions*, Annales Institut Fourier 29 (1979), 239–260.

[3]. *Weighted distribution spaces and pseudodifferential operators*, J. d'Analyse Math. 39 (1981), 131–187.

R. Beals and C. Fefferman

[1]. *Spatially inhomogeneous pseudodifferential operators*, Comm. Pure Appl. Math. 27 (1974), 1–24.

R. Beals, B. Gaveau, P. C. Greiner, and J. Vauthier

[1]. *The Laguerre Calculus on the Heisenberg group, II*, Bull. Sci. Math. 110 (1986), 225–288.

R. Beals and P. C. Greiner

[1]. *Pseudodifferential operators associated to hyperplane bundles*, Rend. Semin. Matem. di Torino (1983), 7–40.

R. Beals, P. C. Greiner and N. K. Stanton

[1]. *The heat equation and geometry of CR manifolds*, Bull. Amer. Math. Soc. 10 (1984), 275–276.

[2]. *The heat equation on a CR manifold*, J. Diff. Geom. 20 (1984), 343–387.

[3]. $L^p$ *and Lipschitz estimates for the* $\bar{\partial}$*-equation and the* $\bar{\partial}$*-Neumann problem*, Math. Ann. 277 (1987), 185–196.

R. Beals, P. C. Greiner and J. Vauthier

[1]. *The Laguerre calculus on the Heisenberg group.* in R. A. Askey et al.,
"Special Functions: Group Theoretical Aspects and Applications,"
Reidel, Dordrecht, 1984, pp. 189–216.

R. Beals and N. K. Stanton

[1]. *The heat equation for the $\bar{\partial}$-Neumann problem, I,* Comm. in P.D.E.
12 (1987), 351–413.

L. Boutet de Monvel

[1]. *Hypoelliptic operators with double characteristics and related
pseudodifferential operators,* Comm. Pure Appl. Math. 27 (1974),
585–639.

L. Boutet de Monvel, A. Grigis and B. Helffer

[1]. *Parametrices d'opérateurs pseudodifferentiels à caracteristiques mul-
tiples,* Astérisque 34–35 (1976), 93–121.

L. Boutet de Monvel and J. Sjöstrand

[1]. *Sur la singularité des noyaux de Bergman et de Szegö,* Astérisque
34–35 (1976), 123–164.

L. Boutet de Monvel and F. Treves

[1]. *On a class of pseudodifferential operators with double characteristics,*
Inventiones Math. 24 (1974), 1–34.

A. P. Calderón and R. Vaillancourt

[1]. *A class of bounded pseudodifferential operators,* Proc. Nat. Acad. Sci.
U.S.A. 69 (1972), 1185–1187.

D. C. Chang

[1]. On $L^p$ and Hölder estimates for the $\bar{\partial}$- Neumann problem on strongly
pseudo-convex domains, Ph.D. Dissertation, Princeton 1987.

R. Coifman and G. Weiss

[1]. "Analyse Harmonique non-commutative sur certains espaces
homogènes," Lecture Notes in Math. no. 242, Springer, Berlin,
1971.

A. Dynin

[1]. *An algebra of pseudodifferential operators on the Heisenberg group: symbolic calculus*, Sov. Math. Doklady 17 (1976), 508–512.

[2]. *Pseudodifferential operators on Heisenberg groups*, C.I.M.E. Seminar, Bressanone, Italy, 1977.

Yu. V. Egorov

[1]. *On canonical transformations of pseudo-differential operators*, Uspehi Matem. Nauk 24 (1969), 235–236.

C. Fefferman

[1]. *The Bergman kernel and biholomorphic mappings of pseudo-convex domains*, Inventiones Math. 26 (1974), 1–66.

G. B. Folland

[1]. *A fundamental solution for a subelliptic operator*, Bull. Amer. Math. Soc. 79 (1973), 373–376.

G. B. Folland and J. J. Kohn

[1]. "The Neumann Problem for the Cauchy–Riemann Complex," Annals Math. Studies no. 75, Princeton Univ. Press, Princeton, 1972.

G. B. Folland and E. M. Stein

[1]. *Estimates for the $\overline{\partial}_b$-complex and analysis on the Heisenberg group*, Comm. Pure Appl. Math. 27 (1974), 429–522.

B. Gaveau

[1]. *Principe de moindre action, propagation de la chaleur et estimées sous elliptiques sur certains groupes nilpotents*, Acta Math. 139 (1977), 95–153.

P. Gilkey

[1]. *Curvature and the eigenvalues of the Laplacian for elliptic complexes*, Advances Math. 10 (1973), 344–382.

G. Giraud

[1]. *Sur certaines opérations du type elliptique*, C. R. Acad. Sci. Paris 200 (1935), 1651–1653.

P. C. Greiner

[1]. *An asymptotic expansion for the heat equation*, Archive Rational Mech. Anal. 41 (1971), 163–218.

[2]. *On the Laguerre calculus of left-invariant convolution (pseudodifferential) operators on the Heisenberg group*, Sém. Goulaouic–Meyer–Schwartz 1980–1981, exp. 11, Ecole polytechnique, Palaiseau, 1981.

[3]. *Imbedding $C^n$ in $H_n$*, in "Pseudodifferential Operators and Applications," Proc. Symp. Pure Math. vol. 43, Amer. Math. Soc. Providence, 1985, 133–147.

P. C. Greiner, J. J. Kohn and E. M. Stein

[1]. *Necessary and sufficient conditions for the solvability of the Lewy equation*, Proc. Nat. Acad. Sci. U.S.A. 72 (1975), 3287–3289.

P. C. Greiner and E. M. Stein

[1]. *On the solvability of some differential operators of type $\Box_b$*, Proc. Seminar on Several Complex Variables, Cortona, Italy, 1976–1977, 106–165.

[2]. "Estimates for the $\bar\partial$-Neumann Problem," Mathematical Notes 19, Princeton Univ. Press, Princeton, 1977.

A. Grigis

[1]. *Hypoellipticité et paramétrixes pour des opŕateurs pseudo-différentiels à caracteristiques doubles*, Astérisque 34–35 (1976), 183–205.

V. V. Grušin

[1]. *On a class of hypoelliptic operators*, Mat. Sbornik 83 (1970), 456–473; Math. USSR Sbornik 12 (1970),458–476.

[2]. *On a class of hypoelliptic operators degenerating on a manifold*, Mat. Sbornik 84 (1971), 163–195; Math. USSR Sbornik 13 (1971), 155–185.

L. Hörmander

[1]. *Pseudodifferential operators and hypoelliptic equations*, in "Singular Integrals," Proc. Symp. Pure Math. vol. 10, American Math. Soc., Providence, 1967, 138–183.

[2]. *A class of pseudodifferential operators with double characteristics*, Math. Ann. 217 (1975), 165–188.

A. Hulanicki

[1]. *The distribution of energy in the Brownian motion in the Gaussian field and analytic-hypoellipticity of certain subelliptic operators on the Heisenberg group*, Studia Math. 56 (1976), 165–173.

D. S. Jerison

[1]. *The Dirichlet problem for the Kohn Laplacian on the Heisenberg group, I*, J. Functional Analysis 43 (1981), 97–142.

N. Kerzman and E. M. Stein

[1]. *The Szegö kernel in terms of Cauchy–Fantappié kernels*, Duke Math. J. 45 (1978), 197–224.

J. J. Kohn

[1]. *Boundaries of complex manifolds*, in Proc. Conference on Complex Analysis, Minneapolis 1965, Springer, Berlin, 1965, 81–93.

J. J. Kohn and L. Nirenberg

(1) *An algebra of pseudo-differential operators*, Comm. Pure Appl. Math. 18 (1965), 269–305.

J. J. Kohn and H. Rossi

[1]. *On the extension of holomorphic functions from the boundary of a complex manifold*, Annals Math. 81 (1965), 451–472.

A. Korányi and S. Vági

[1]. *Singular integrals in homogeneous spaces and some problems of classical analysis*, Annali Scuola Norm. Sup. Pisa 25 (1971), 575–648.

H. Lewy

[1]. *On the local character of the solutions of an atypical linear differential equation in three variable and a related theorem for regular functions of two complex variables*, Annals Math. 64 (1956), 514–522.

[2]. *An example of a smooth linear partial differential equation without solution*, Ann. Math. 66 (1957), 155–158.

H. P. McKean, Jr. and I. M. Singer

[1]. *Curvature and the eigenvalues of the Laplacian*, J. Diff. Geom. 1 (1967), 43–69.

A. Melin

[1]. *Lie filtrations and pseudodifferential operators*, preprint.

A. Menikoff

[1]. *Subelliptic estimates for pseudodifferential operators with double characterics*, mimeographed, 1975.

S. G. Mikhlin

[1]. *Compounding of double singular integals*, Dokl. Akad. Nauk SSSR 2 (11), (1936), 3–6.

A. Nagel and E. M. Stein

[1]. *A new class of pseudodifferential operators*, Proc. Nat. Acad. Sci. U.S.A. 75 (1978), 582–585.
[2]. "Lectures on Pseudo-differential Operators: Regularity Theorems and Applications to Non-Elliptic Problems," Mathematical Notes, Princeton Univ. Press, Princeton, 1979.

L. Nirenberg

[1]. *Pseudo-differential operators*, in "Global Analysis," Proc. Symp. Pure Math. vol. 16, Amer. Math. Soc., Providence, 1970, 147–168.
[2]. *On a problem of Hans Lewy*, Uspehi Mat. Nauk 292 (176) (1974), 241–251; Lecture Notes in Math. no. 459, Springer, Berlin, 1975, 224–234.

V. K. Patodi

[1]. *Curvature and the eigenforms of the Laplace operator*, J. Diff. Geom. 5 (1971), 233–249.

D. H. Phong nd E. M. Stein

[1]. *Estimates for the Bergman and Szegö projections on strongly pseudo-convex domains*, Duke Math. J. 44 (1977), 695–704.

L. P. Rothschild and E. M. Stein

[1]. *Hypoelliptic differential oprators and nilpotent groups*, Acta Math. 137 (1976), 247–320.

L. P. Rothschild and D. S. Tartakoff

[1]. *Parametrices with $C^\infty$ error for $\square_b$ and operators of Hörmander type*, "Partial Differential Equations and Geometry," Marcel Dekker, New York, 1979, 255–271.

R. T. Seeley

[1]. *Refinement of the functional calculus of Calderón and Zygmund*, Proc. Akad. Wet. Ned. Ser. A 68 (1965), 521–531.
[2]. *Complex powers of an elliptic operator*, in "Singular Integrals," Proc. Symp. Pure Math. vol. 10, Amer. Math. Soc., Providence, 1967, 288–307.
[3]. *Elliptic singular integral operators*, in "Singular Integrals," Proc. Symp. Pure Math. vol. 10, Amer. Math. Soc., Providence, 1967, 308–315.

J. Sjöstrand

[1]. *Parametrices for pseudodifferential operators with multiple characteristics*, Arkiv för Math. 12 (1974), 85–130.

N. K. Stanton and D. S. Tartakoff

[1]. *The heat equation for the $\overline{\partial}_b$- Laplacian*, Comm. P.D.E. 9 (1984), 597–686.

G. Szegö

[1]. "Orthogonal Polynomials," Colloq. Publ. vol 23, Amer. Math. Soc., Providence, 1939.

D. S. Tartakoff

[1]. *Local Gevrey and quasi-analytic hypoellipticity for $\square_b$*, Bull. Amer. Math. Soc. 82 (1976), 740–742.
[2]. *Local analytic hypoellipticity of $\square_b$ on nondegenerate Cauchy-Riemann manifolds*, Proc. Nat. Acad. Sci. U.S.A. 75 (1978), 3027–3028.
[3]. *The local real analyticity of solutions to $\square_b$ and the $\overline{\partial}$-Neumann problem*, Acta Math. 145 (1980), 177–204.

M. E. Taylor

[1]. "Pseudodifferential Operators," Princeton Univ. Press, Princeton, 1981.

[2]. "Noncommutative Microlocal Analysis, Part I," Memoirs Amer. Math. Soc. no. 313, Providence, 1984.

F. Treves

[1]. *Analytic hypoellipticity for a class of pseudodifferential operators*, Comm. P.D.E. 3 (1978), 475–642.
[2]. "Introduction to Pseudodifferential and Fourier Integral Operators," vol. I., Plenum, New York, 1980.

A. Unterberger and J. Bokobza

[1]. *Des Opérateurs de Calderón–Zygmund précisés*, C.R. Acad. Sci. Paris, sér. A, 260 (1965),3265–3267.

# Index of Terminology

# List of Notation

$\mathcal{G}_k$, 128

$H^m(\mathbf{R}^n)$, $H^m_{\text{loc}}(U)$, 86

$\mathcal{K}_j$, 130

$\mathcal{K}^m$, 136

$N_{b,q}$, 169

$P_\lambda$, 27

$P^y$, 12

$[Q]$, 147

$Q^y$, 104

$S^\infty$, 24

$S^m(U)$, 80

$S^{-\infty}(U)$, $S^\infty(U)$, 81

$S^m_{\frac{1}{2},\frac{1}{2}}(U)$, 88

$S_{m,\nu}(U)$, 91

$S^m_\nu(U)$, 91

$S^m_y$, 106

$T_{0,1}$, $T_{1,0}$, 151

$\#$ -composition, 116, 118